COASTAL METEOROLOGY

A REVIEW OF THE STATE OF THE SCIENCE

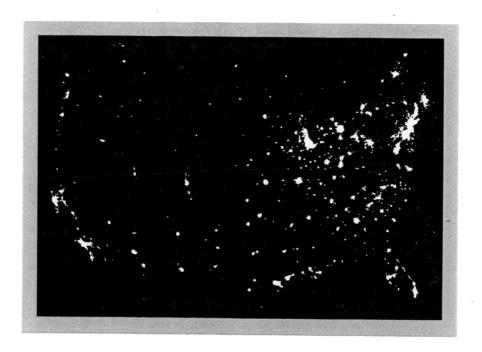

Panel on Coastal Meteorology
Committee on Meteorological Analysis, Prediction, and Research
Board on Atmospheric Sciences and Climate
Commission on Geosciences, Environment, and Resources
National Research Council

NATIONAL ACADEMY PRESS
Washington, D.C. 1992

NOTICE: The project that is the subject of this report was approved by the Governing Board of the National Research Council, whose members are drawn from the councils of the National Academy of Sciences, the National Academy of Engineering, and the Institute of Medicine.

This report has been reviewed by a group other than the authors according to procedures approved by a Report Review Committee consisting of members of the National Academy of Sciences, the National Academy of Engineering, and the Institute of Medicine.

The National Academy of Sciences is a private, nonprofit, self-perpetuating society of distinguished scholars engaged in scientific and engineering research, dedicated to the furtherance of science and technology and to their use for the general welfare. Upon the authority of the charter granted to it by the Congress in 1863, the Academy has a mandate that requires it to advise the federal government on scientific and technical matters. Dr. Frank Press is president of the National Academy of Sciences.

The National Academy of Engineering was established in 1964, under the charter of the National Academy of Sciences, as a parallel organization of outstanding engineers. It is autonomous in its administration and in the selection of its members, sharing with the National Academy of Sciences the responsibility for advising the federal government. The National Academy of Engineering also sponsors engineering programs aimed at meeting national needs, encourages education and research, and recognizes the superior achievements of engineers. Dr. Robert M. White is president of the National Academy of Engineering.

The Institute of Medicine was established in 1970 by the National Academy of Sciences to secure the services of eminent members of appropriate professions in the examination of policy matters pertaining to the health of the public. The Institute acts under the responsibility given to the National Academy of Sciences by its congressional charter to be an adviser to the federal government and, upon its own initiative, to identify issues of medical care, research, and education. Dr. Kenneth I. Shine is president of the Institute of Medicine.

The National Research Council was organized by the National Academy of Sciences in 1916 to associate the broad community of science and technology with the Academy's purposes of furthering knowledge and advising the federal government. Functioning in accordance with general policies determined by the Academy, the Council has become the principal operating agency of both the National Academy of Sciences and the National Academy of Engineering in providing services to the government, the public, and the scientific and engineering communities. The Council is administered jointly by both Academies and the Institute of Medicine. Dr. Frank Press and Dr. Robert M. White are chairman and vice chairman, respectively, of the National Research Council.

Support for this project was provided jointly by the Department of the Interior, the National Oceanic and Atmospheric Administration, the National Science Foundation, the U.S. Army Research Office, the U.S. Army Atmospheric Sciences Laboratory, the U.S. Army Corps of Engineers (Waterways Experiment Station), the U.S. Coast Guard (R&D Center), and the U.S. Navy Office of Naval Research under Grant No. N00014-90-J-4138.

Copies of this report are available from: **NATIONAL ACADEMY PRESS, 2101 Constitution Avenue, N.W., Washington, DC 20418**

Printed in the United States of America

Cover photo: The United States at night: From 250 miles above the earth, lights of cities and towns outline densely populated coasts. Complex weather created by adjoining water and land surfaces affects over 100 million people in the United States. Photo courtesy of the Air Force Global Weather Central and the Defense Meteorological Satellite Program.

Preface

The unique weather and climate of the coastal zone, where the very different properties of land and sea meet, strongly affect pollutant circulation, storm characteristics, air and sea current patterns, and local temperatures. Nearly half the U.S. population currently lives in coastal areas,[1] and this number is expected to grow in the next 20 years from about 110 million to more than 127 million people. A better understanding of coastal meteorology would thus be of considerable benefit to the nation, since it affects air pollution and disaster preparedness; ocean pollution and safeguarding near-shore ecosystems; offshore oil exploration and drilling; military and merchant ship operations; and a host of other activities affecting commerce, industry, transportation, health, safety, recreation, and national defense.

As a result of progress in several areas of meteorological research, as well as the development of new technologies, opportunities now exist for significant advances in both basic understanding and forecasting of a wide variety of important coastal meteorological phenomena. In recent years new in situ and remote sensing measuring techniques have become available that can be used to study and monitor coastal phenomena in considerable detail. Numerical models are now available with sufficiently small grid spacings to resolve many coastal meteorological events. Widespread availability of small but powerful computer workstations will permit both

---

[1]Department of Commerce (1990), *Fifty Years of Population Change Along the Nation's Coasts, 1960-2010,* National Ocean Service, National Oceanic and Atmospheric Administration, Washington, D.C., 41 pp.

research studies and operational forecasting of important weather phenomena along coastlines, many of which depend on specific aspects of local geography and topology.

This report reviews the progress that has been made in recent years by the small research community engaged in studies of coastal meteorology. It is intended to guide researchers into those areas in which their efforts might be most productive. It should also alert policy makers, local and federal authorities, and private organizations to the new tools that are available for improving the safety and efficiency of operating in and managing coastal regions.

Following a general introduction to the subject, this report reviews recent progress and current understanding of coastal meteorological phenomena, including land and sea breezes, coastal fronts, orographic effects, landfalling hurricanes, air quality, and coastal effects in the polar regions. Gaps in knowledge are identified, and recommendations for advancing basic understanding and applications are given at the end of each chapter. Final chapters address educational and human resource issues and highlight the new observational and modeling tools that can be brought to bear on coastal meteorological research and operations.

On behalf of the Committee on Meteorological Analysis, Prediction, and Research, I wish to thank the panel members, particularly the panel's chairman, Richard Rotunno, for the outstanding job they have done in producing a report of value to both scientists and policy makers. Thanks are extended to Alan Weinstein of the Office of Naval Research for having the foresight to suggest this study and for recognizing the broad applications and interests in coastal meteorology among several federal agencies. Early collaboration with the Committee on the Coastal Ocean of the National Research Council's Ocean Studies Board in helping to form the panel is gratefully acknowledged. Thanks are also extended to John S. Perry and Kenneth Bergman for initial staff support of the study and to William A. Sprigg for helping to guide the report to its completion.

Peter V. Hobbs, *Chairman*
Committee on Meteorological Analysis,
Prediction, and Research

Contents

EXECUTIVE SUMMARY   1

1 INTRODUCTION   5

2 BOUNDARY LAYER PROCESSES   9
    Current Understanding and Challenges, 10
      The Generic Atmospheric Boundary Layer, 10
      Surface Interactions, 13
      Internal Boundary Layers, 15
      The Inhomogeneous Atmospheric Boundary Layer, 16
      Boundary Layer Clouds, 17
    Summary and Conclusions, 18

3 THERMALLY DRIVEN EFFECTS   19
    The Land Breeze and the Sea Breeze, 19
    Coastal Fronts, 26
    Ice-Edge Boundaries, 27
    Summary and Conclusions, 29

4 THE INFLUENCE OF OROGRAPHY   31
    Introduction and Basic Parameters, 31
    Low Froude Number Flow: Trapped Phenomena, 33
      Isolated Response: Kelvin Wave
        and Gravity Current, 33

Damming, 35
Gap Winds, 36
Moderate Froude Number Flow, 38
Katabatic and Other Local Winds, 39
Summary and Conclusions, 43

5 INTERACTIONS WITH LARGER-SCALE
   WEATHER SYSTEMS   45
   Land-Falling Hurricanes, 45
   Polar and Arctic Lows, 46
   Hybrid Frontal Circulations and Winter Storms, 47
   Localized Latent Heat Release, 48
   Summary and Conclusions, 49

6 THE INFLUENCE OF THE ATMOSPHERIC BOUNDARY
   LAYER ON THE COASTAL OCEAN   51
   Coastal Processes, 53
     Local and Remote Wind Forcing, 53
     Ocean Fronts, 54
     Storms, 55
     Topography, 56
   Air-Sea Exchange Processes, 57
     Wind Stress, Heat Fluxes, and Trace Gas and
       Particulate Exchange, 57
     Coupled Interactions with the Planetary
       Boundary Layer, 58
   Summary and Conclusions, 59

7 AIR QUALITY   63
   Dispersion in the Coastal Zone, 63
   Developing Improved Dispersion Models, 67
   Summary and Conclusions, 68

8 CAPABILITIES AND OPPORTUNITIES   71
   Observational Tools, 71
     In Situ Methods, 71
     Remote Sensing, 74
   Modeling Technology, 78

9 EDUCATIONAL AND HUMAN RESOURCES   81

REFERENCES   85

COASTAL METEOROLOGY

A Review of the State of the Science

Executive Summary

Coastal meteorology is the study of meteorological phenomena in the coastal zone—that is, within about 100 km inland or offshore of a coastline. Weather in this region is caused, or significantly affected, by the sharp changes that occur between land and sea in surface transfers and/or elevation. With regard to surface transfers, study of the atmospheric boundary layer (ABL) is fundamental. However, existing understanding of the ABL is applicable primarily to horizontally homogeneous conditions; it is therefore poorly suited to the coastal zone, a region of strong horizontal inhomogeneity. Hence, future observations and theories must focus more on the horizontally inhomogeneous ABL.

Differences in the vertical heat transfer across a coastline play a role in a number of coastal meteorological phenomena, such as the land-sea breeze and coastal fronts. Although these circulations are roughly understood, a deeper understanding is needed to make accurate predictions on the meso-scale. Further progress will come with high-density observational and high-resolution numerical modeling studies of situations with curved coastlines, heterogeneous surfaces, time-dependent large-scale flow, and clouds.

Changes in elevation across a coastline significantly affect coastal meteorology. In many situations the coastal mountains act as a barrier to the stably stratified marine air; the barrier may block air flowing toward it, or it may act like a wall along which Kelvin waves may propagate. Further study is needed of the ageostrophic dynamics of these and other mesoscale wind features.

Interactions of larger-scale weather systems with the coastal environment frequently involve a number of processes already mentioned, and their nature is often difficult to distinguish in complex situations. The panel believes that focusing on specific interactions—such as clouds interacting with sea breezes, mountains interacting with coastal winds, and the effect of coastal fronts on extratropical cyclogenesis—will lead to better understanding and hence to applications such as air quality and pollutant dispersal modeling. Researchers may begin to concentrate on developing improved techniques for mesoscale data assimilation and to pave the way for site-specific forecasting of coastal weather and sea state.

In the area of air-sea interactions in the coastal zone, the local processes governing air-sea fluxes within an inhomogeneous boundary layer and variable wave state need to be better understood, as does the role of mesoscale spatial inhomogeneities in controlling coastal dynamics.

The panel found that existing buoy and coastal station networks are outdated and inadequate in number (too often failing in accuracy and precision of measurement), especially for obtaining data over water. The panel also encourages collaboration between meteorologists and oceanographers through research programs and enhancements in college-level curricula that focus on problems of coastal meteorology and oceanography.

The chapters that follow detail these findings and suggest a number of specific courses of action. The panel has drawn from these suggestions the following general recommendations:

With respect to boundary layer processes, WE RECOMMEND a complete reexamination of ABL processes in inhomogeneous conditions, including surface and boundary layer scaling theories, higher-order moment relationships throughout the ABL, and the relative importance of turbulent versus coherent motions.

To enhance knowledge of thermally driven effects in coastal regions, WE RECOMMEND that high-density land and sea breeze observational studies and high-resolution numerical modeling studies in regional field experiments be extended to three dimensions using complex coastlines and topography, heterogeneous surfaces, and nonhomogeneous and time-dependent synoptic environments that include cloud interactions. WE RECOMMEND also that detailed study be undertaken of other thermally driven circulations and influences, nondiurnal in nature, including persistent coastal fronts and phenomena associated with ice sheet leads and polynyas.

To develop greater understanding of orographic influences on coastal meteorological phenomena, WE RECOMMEND numerical and observational investigations of ageostrophic dynamics for the initiation, intensification, and movement of coastal mesoscale wind features, such as coastal jets and eddies and gap winds, caused by the interaction of the synoptic-scale flow with coastal orography.

To understand the nature of the interactions of large-scale weather systems with the coastal environment, WE RECOMMEND observational, numerical, and theoretical studies that focus on specific interactions—for example, the effect of coastal fronts on extratropical cyclogenesis—in order to develop an understanding of the dynamical processes involved.

To improve understanding of the influence of the ABL on the coastal ocean, WE RECOMMEND that a research program be undertaken to clarify (1) the local physical and chemical processes governing air-sea fluxes of momentum, heat, moisture, particulates, and gases within an inhomogeneous coastal boundary layer and variable wave state and (2) the role of distant mesoscale spatial inhomogeneities in controlling atmosphere-ocean dynamics in a coastal environment.

To address air quality issues in coastal regions, WE RECOMMEND use of advanced modeling systems and tracer tests (for verification) to determine the significant impacts of vertical motions and shears in three-dimensional coherent mesoscale coastal circulations on the dispersion of gases, aerosols, and particulates, especially in the range of 10 to 100 km.

To apply advanced technology in coastal research, WE RECOMMEND (1) the use of recently developed remote sensors to obtain detailed four-dimensional data sets along with the upgrading of buoy and surface station networks to obtain quality, long-duration data sets describing coastal regions and (2) on-site use of high-performance workstations to provide decentralized computations during study of local coastal phenomena, data assimilation methods, and real-time forecasting.

Finally, to focus attention on the subject of coastal meteorology, WE RECOMMEND increased use of interdisciplinary conferences and short courses, together with support of university training programs, to encourage more scientists to explore the meteorology of the coastal zone.

1

Introduction

Coastal meteorology is the study of meteorological phenomena in the coastal zone caused, or significantly affected, by sharp changes in heat, moisture, and momentum transfers and elevation that occur between land and water. The coastal zone is defined as extending approximately 100 km to either side of the coastline. Examples of coastal meteorological phenomena include land and sea breezes, sea-breeze-related thunderstorms, coastal fronts, fog, haze, marine stratus clouds, and strong winds associated with coastal orography. In addition to their intrinsic importance to coastal weather, increased knowledge of these phenomena is important for understanding the physical, chemical, and biological oceanography of the coastal ocean. Practical application of this knowledge is vital for more accurate prediction of coastal weather and sea states, which affect transportation and commerce, pollutant dispersal, public safety, and military operations. This report reviews the state of the science of coastal meteorology. In addition, it recommends areas for scientific and technical progress.

The dynamical meteorology of the coastal zone may be considered in terms of the three subsidiary ideal problems illustrated in Figure 1.1; these three problems form the organizational basis for this report. The first problem is one where the coastal atmospheric circulation is primarily driven by the contrast in heating and is modulated by the contrast in surface friction between land and water. The second problem is one where the primary influence is due to steep coastal mountains, the presence of which may induce strong winds and other complex flow patterns. A third class of phenomena broadly consists of larger-scale meteorological systems that, by

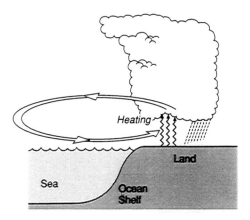

Thermally Driven Effects

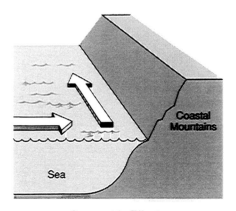

Orographic Effects

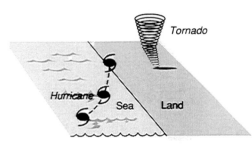

Interaction with Larger-Scale
Weather Systems

FIGURE 1.1   Three categories of problems in coastal meteorology.

virtue of their passage across the coastline, produce distinct smaller-scale systems. Real coastal phenomena are always some combination of these idealized problems.

Transfers of heat, momentum, and water vapor between the atmosphere and its underlying surface (be it land or water) are basic to these three ideal problems. This report therefore begins with *Boundary Layer Processes* (Chapter 2); this chapter contains an assessment of, and prospects for improvement in, our understanding of the approximately 1-km-deep layer of air adjacent to the surface, which is called the atmospheric boundary layer (ABL). Study of the ABL is intended to reveal how surface transfers are distributed upward. Over the ocean, those surface transfers are interactive, determined by the sea state, which in turn is determined by the atmospheric flow, which is influenced by the surface transfers, and so on. This fundamental coupling has been long recognized. However, there is another order of complexity over the coastal ocean because the sea state is significantly influenced by the ocean bottom. Over land, there is still significant uncertainty about the nature of surface transfer from terrain with variations in soil moisture, vegetation, and usage such as occur along the coast. These strong and frequently horizontal variations in surface transfers form a particularly formidable impediment to understanding of the ABL in coastal regions. But even in the absence of such horizontal variations, the marine boundary layer containing stratus clouds and drizzle is a complex problem involving the interplay of turbulence, cloud processes, and radiation.

Problems in the first general category are discussed in *Thermally Driven Effects* (Chapter 3). Although recognition of the land-sea breeze dates back to antiquity, the understanding needed to make accurate forecasts is still lacking. In simple terms, the land-sea breeze is caused by the generally different temperatures of the land and sea, which produce an across-coast air temperature contrast. After this circulation begins, however, it modifies the conditions that produced it. Thus, the difficulty in making precise predictions lies in understanding more clearly the nature of this feedback. Uncertainties in our understanding of the ABL and nonlinear feedbacks between the sea breeze and resultant clouds are examples here. Issues associated with two special types of thermally driven phenomena (coastal fronts and ice-edge boundaries) also are discussed in Chapter 3.

Coastal mountain ranges can significantly affect coastal meteorology. In *The Influence of Orography* (Chapter 4), the types of problems encountered are discussed. In many situations, coastal mountains act as a barrier to the stably stratified marine air. Thus, air with an initial component of motion toward the barrier must eventually turn and flow along the barrier. Under the influence of the earth's rotation, waves known as Kelvin waves (see Figure 1.1) can propagate along a basin-wall-like coastal mountain range. This type of motion is an important component of the meteorological problems in these regions. Local effects such as katabatic winds, gap winds, and eddies also are discussed in Chapter 4.

As larger-scale meteorological systems move across the coast, they are affected by some combination of the effects discussed in the previous two paragraphs. In some situations, distinct subsystems, which would not exist without the coastal influence, are produced. These are discussed in *Interactions with Larger-Scale Weather Systems* (Chapter 5). Examples of these effects include cyclogenesis, which is enhanced at the east coast of the United States as upper-level disturbances cross the Appalachians and encounter the strong baroclinic zone at the coast; flow along the coast in winter with strong cooling of the air on the landward side, leading to the formation of fronts; and land-falling hurricanes whose low-level flows are so modified as to favor the formation of tornadoes.

In general, the ocean affects, and is affected by, the atmosphere. In *The Influence of the Atmospheric Boundary Layer on the Coastal Ocean* (Chapter 6), aspects of this interaction that are particularly important for the ocean part of the coastal zone (shelf waters) are discussed. The processes governing air-sea fluxes of momentum, heat, mass, and gases are described in terms of local and remote forcings as well as wave state. Wind-driven coastal upwelling of colder water from below brings different chemical and biological compositions to the surface, produces an across-coast temperature difference unique to coastal regions, and influences atmospheric circulation. Interactions of this nature are important to understanding the coastal ocean and the chemical and biological processes occurring there.

Another important application of coastal meteorology is the prediction of pollutant dispersal. In *Air Quality* (Chapter 7), the focus is on physical advective processes. The highly variable winds near the coast may sweep pollutants out to sea on a land breeze, but then bring them back with the sea breeze. More accurate estimates of the vertical motion fields associated with these wind systems are critical to determining the layers in which the pollutant resides (and the horizontal direction in which it will move).

Capabilities and Opportunities (Chapter 8) discusses new observational and simulation tools that can be exploited to study the coastal environment. The critical issue is to have instruments in place to measure long time-scale variations (from 1 or 2 days to a season), as well as short-time fluctuations, so that relationships between the two can be discerned. The use of remote sensing as well as in situ devices is discussed in this context. Increases in the computer power of desktop machines will allow many investigators to explore, with numerical models, flow in the vicinity of many coastal environments. With regard to *Educational and Human Resources* (Chapter 9), the panel found that there is a relative lack of training of meteorology students in areas pertaining to coastal meteorology and a relative lack of cross-fertilization between the fields of coastal meteorology and coastal oceanography.

2

Boundary Layer Processes

The atmospheric boundary layer (ABL) is the part of the lower troposphere that interacts directly with the earth's surface through turbulent transport processes. A coast separates two drastically different surfaces, and a coastal region has an inhomogeneous boundary layer. ABL processes are important in determining the evolution of atmospheric structure. The boundary layer is also a buffer zone that interacts both with the "free" tropospheric flow at its upper interface (through entrainment processes) and with the surface (through surface exchange processes).

Past studies of the ABL have emphasized certain idealized, near-equilibrium, and horizontally homogeneous boundary layer regimes (Wyngaard, 1988). For example, Stull's (1988) comprehensive reference on boundary layer meteorology devotes only about 5 percent of its discussion to geographic effects. In fact, many common ABL conditions are still poorly understood even for homogeneous surfaces. The horizontally inhomogeneous and rapid temporal forcing conditions typical of coastal regions dictate consideration of problems that have rarely been investigated. Furthermore, even those aspects of the physical processes that are generally regarded as being well understood (e.g., bulk parameterization of surface fluxes) must be reevaluated before applying them to coastal environments.

In this chapter, current understanding of boundary layer processes is examined, and some important deficiencies are identified. Following introductory material on boundary layers, generic problems in boundary layer processes are examined. Then, special coastal problems are considered that involve surface fluxes, internal boundary layer growth, baroclinicity, and a

9

variety of phenomena that are either inherently inhomogeneous or are associated with inhomogeneous forcing.

CURRENT UNDERSTANDING AND CHALLENGES

The ABL has been investigated extensively. Certain aspects are considered to be well understood. This understanding has developed from a combination of sources: laboratory models (e.g., Deardorff and Willis, 1982; Willis and Deardorff, 1974); three-dimensional primitive equation large-eddy simulations (e.g., Deardorff, 1974, 1980; Moeng, 1984a, b); and atmospheric measurements with aircraft and tethered balloons (e.g., Brost et al., 1982; Kaimal et al., 1976; Lenschow, 1973; Lenschow et al., 1980; Nicholls, 1984).

A variety of ABL models are now available. Because a model represents the reduction of a problem to its important components, it not only is a useful tool but also an expression of our understanding of the physics. Linear regression and crude parameterizations often imply little or no understanding. Today, a hierarchy of complexity is available in atmospheric models. Similarity models are the simplest but typically are applicable only to idealized situations. One index of understanding is in simplified conceptual models and useful scaling laws. By this measure, the cloud-free convective ABL is clearly the best-understood regime.

Numerical solutions to systems of physical equations form the basis of the more sophisticated models commonly used in meteorology. Two approaches are used, depending on whether solutions are sought for the ensemble average or for the volume average atmospheric budget and state equations. Ensemble average models are often referred to as higher-order closure models. Grid-volume average models are usually referred to as large eddy simulation (LES) models. These approaches are fundamentally different; an LES model produces an explicit simulation of a single realization of a three-dimensional, time-dependent atmospheric structure. An ensemble average model predicts or describes relationships between the moments of the atmospheric variables (the first moments are averages of the variables and the second moments are variances and fluxes). LES models are currently used strictly for research (such as developing parameterizations); ensemble average models have a variety of practical as well as research applications.

The Generic Atmospheric Boundary Layer

Scaling theories have their origins in dimensional analysis; important variables of the problem are selected, and other properties are calculated from dimensionally consistent combinations of those variables. Modern

ABL similarity theories are now based on arguments about the relative variability and magnitude of the terms in the mean and turbulent budget equations. In the ABL the similarity regimes are broken down by the vertical scale, assuming that the horizontal fields are statistically homogeneous. Historically, this process has proceeded from the ground up. Figure 2.1 depicts schematically the idealized mean structure of the ABL under convective, well-mixed, cloud-free, horizontally homogeneous conditions in a synoptic regime that has sufficient subsidence to ensure the presence of a capping inversion. Compared to the ABL, the overlying free troposphere can be considered essentially nonturbulent.

Surface layer similarity theory is based on scaling parameters obtained from the surface fluxes (Wyngaard, 1973). The theory is considered valid in the region near the surface where various terms (particularly the gradients) in the turbulent kinetic energy and scalar variance budget equations are considerably more dependent on height than are the fluxes. Thus, the assumptions on which the theory is based are generally valid in the lowest 10 percent of the ABL. For the convective ABL, we also have mixed layer similarity (Moeng and Wyngaard, 1989) and inversion region similarity (Wyngaard and LeMone, 1980). For the stable ABL, no mixed layer exists. Instead, there is a gradual transition from the surface layer to the inversion layer. A local similarity theory (Nieuwstadt, 1984) has been proposed for a

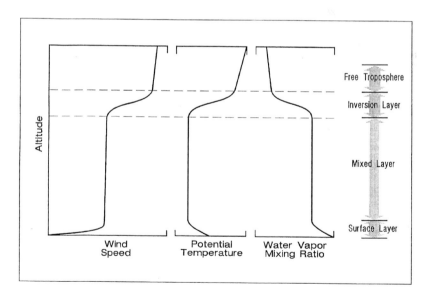

FIGURE 2.1 Idealized profiles of wind speed, potential temperature, and water vapor mixing ratio for the convectively mixed, cloud-free atmospheric boundary layer (after Fairall et al., 1982).

stable ABL in steady-state or slowly evolving conditions. This theory has, however, exhibited shortcomings in general application.

In terms of simple models, the present state of understanding of the generic ABL can be described crudely as follows. It is natural to classify conditions of the ABL by dynamical regimes of increasing complexity: cloud-free, convective; cloud-free and shear-driven; baroclinic; stable; stratocumulus; tradewind cumulus; and broken clouds. Coastal meteorology encompasses all seven of these regimes. Cloud-modified and stable boundary layers remain essentially unsolved problems. A general similarity theory that can handle all possible cloud regimes does not exist. Higher-order closure models have yet to demonstrate detailed agreement for even the simplest cases (Holt and Raman, 1988), and recent LES studies (Moeng and Wyngaard, 1989) have called into question the transport and dissipation closures used in most second-order models. Transport and dispersion properties of the ABL are strongly dependent on the characteristics of coherent structures and the higher-order moments (Weil, 1990). These properties have only recently been studied for the homogeneous convective ABL (Moeng and Rotunno, 1990; Moeng and Wyngaard, 1989). Although some understanding of the idealized ABL (Figure 1.1) exists, actual coastal ABL structure is often significantly different. For example, Figure 2.2 shows a stable

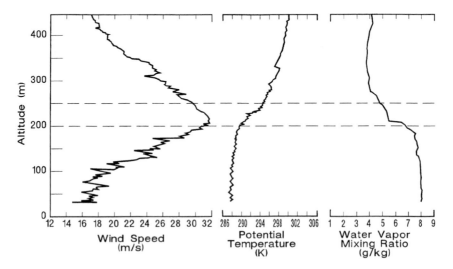

FIGURE 2.2 Vertical profiles of wind speed, potential temperature, and water vapor mixing ratio along the north coast of California (after Winant et al., 1988). Note the strong low-level inversion in potential temperature at approximately 250 m altitude, coincident with the maximum in wind speed. Contrast these profiles with the simple mixed layer structure depicted in Figure 2.1.

ABL with a baroclinically induced jet but a well-mixed humidity. Thus, there is a need both to improve our understanding of the idealized case and to study cases that exhibit stable boundary layers as well as those with strong baroclinicity.

Surface Interactions

Surface fluxes can be measured in homogeneous and moderately inhomogeneous terrain using the eddy correlation technique (Businger, 1986; McMillen, 1988; Wyngaard, 1988), which is considered to be the measurement standard. In numerical models (or when the direct method is not available or applicable), bulk transfer coefficients and surface layer similarity parameterizations are used to relate the fluxes to the near-surface mean meteorological variables and the surface properties. This approach is often quite successful over the open ocean but cannot be straightforwardly applied over land; terrain, soil, and plant canopy interactions greatly complicate the physics (Priestly and Taylor, 1972; Sellers et al., 1986). This is the situation for local coastal climatology where the intensity of the land-sea breeze circulation is relatively much stronger with dry, lightly vegetated coastal lands (Segal et al., 1988).

Inhomogeneous surfaces cause special problems because it is difficult to relate the point measurements used to characterize the surface to the larger-scale mean fluxes (Schuepp et al., 1990). Also, the bulk expressions are intended to relate the average flux to the average bulk variables. Since ABL-scale eddy turnover time is about 15 minutes, it takes about an hour to average ABL variability to obtain a representative sample. In that hour a parcel of air in the ABL can easily travel horizontally about 20 km. This calls into question whether one can rely on surface similarity expressions that are obtained from 1-hour averages of field measurements over homogeneous terrain in models with 1-minute time steps applied over 10×10 km horizontal grids. For example, Beljaars and Holtslag (1991) found that characterization of momentum transfer on horizontal scales of a few kilometers required an "effective" roughness length considerably greater than the local value. The situation for moisture and sensible heat transfer is even more difficult. To quote Beljaars and Holtslag (1991), "More complicated land surface schemes are certainly available to describe the physics in more detail. . . . [H]owever, it is not clear whether all the parameters that specify the land surface in such models can easily be determined." There is also a distinction between a patchy surface that is statistically homogeneous and a "nonstationary" situation where the average properties vary with position and time. The applicability of surface layer similarity and the implications for the bulk transfer coefficients for these conditions are virtually unexplored.

On the ocean side of a coastal region, special problems arise in interfacial transfer. For the open ocean, a reasonable set of bulk transfer coefficients (e.g., Smith, 1988) is available that appears adequate for many applications. These coefficients represent air-sea transfer processes for average surface wave conditions as a function of mean wind speed. Wind speed, wave spectrum, and bulk transfer coefficients have been related theoretically (Geernaert et al., 1986; Huang et al., 1986), but field measurements in coastal regions (e.g., Geernaert et al., 1987; Smith et al., 1990a, b) have demonstrated greatly increased drag coefficients in shallow water (Figure 2.3).

It is clear that fetch and shallow water effects upset the normal equilibrium wind-wave relationships. Fetch is primarily an issue for offshore wind conditions, but distortion of the open ocean directional wave spectrum in coastal shallow regions is important regardless of wind regime; it becomes increasingly important as wind speeds increase. Sounds and estuaries add complexities, often with irregular coastlines, river deltas, and islands. Effects on heat and gas fluxes also are unknown. (See Chapter 6 for further discussion and Geernaert (1990) for a comprehensive review of these con-

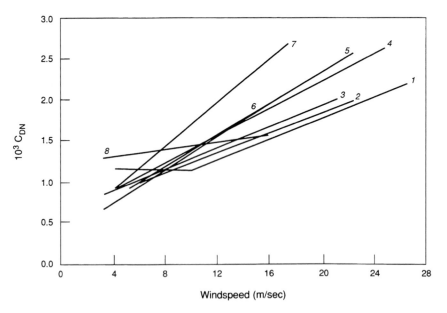

FIGURE 2.3 Distribution of neutral drag coefficient, C_{DN}, with wind speed: (1) over deep open ocean, (2) over deep coastal ocean, (3) over deep water, (4) North Sea depth of 30 m, (5) North Sea depth of 16 m, (6) Lough Neagh depth of 15 m, (7) Lake Ontario depth of 10 m, (8) Lake Geneva depth of 3 m (from Geernaert, 1990).

cepts.) The evolving wave field is influenced by the wind stress, but the stress vector produced by a given wind field is dependent on the directional wave field. Thus, predictions of wind and wave fields on the continental shelves are strongly coupled. Significant and unexplained differences between the mean wind direction and the mean stress direction have been observed in coastal regions (Geernaert, 1990). Substantial hydrostatic stability modulation of stress and surface wind fields has also been observed in association with sea surface temperature variations typical of coastal regions. A comprehensive theoretical and experimental study of wind-wave-stress-scalar flux relationships on the continental shelves should be an important component of a coastal meteorology research program.

Internal Boundary Layers

Air that is modified by flow over an abrupt change in surface properties is said to be confined to an internal boundary layer (IBL). Here we are concerned with the effects of a boundary between two different but individually homogeneous surfaces, as opposed to the boundary layer effects of more general forms of inhomogeneity discussed below. When the surface heat flux changes at such a boundary, a thermal IBL (Garratt, 1987; Lyons, 1975) is formed on the downwind side. Growth in the depth of the IBL is usually parameterized in terms of the downwind distance from the interface. Typically, the form is that of a power law, but many different formulas are available. Convective conditions promote rapid growth; thus, the IBL depth quickly reaches the existing capping inversion. In this case IBL considerations are important only close to the transition. Stable, convection-suppressing conditions downwind of the transition result in slow IBL growth, and the tendency is to form a permanent surface-based inversion (Mulhearn, 1981). The situation is similar to the afternoon-evening transition for the overland convective ABL (Zeman and Lumley, 1979). In this case any turbulence above the new surface-based inversion can be cut off from the surface source of energy and, in the absence of other sources, begin to decay. In an LES study of the decay of convective turbulence, Nieuwstadt and Brost (1986) found that the ABL depth, divided by the convective mixing velocity, formed a characteristic decay time scale, but the behavior of various turbulent variables was not easily parameterized.

Currently, the standard approach to describing near-surface meteorological profiles is to use surface similarity expressions with one set of scaling parameters for the IBL and a second set for the old ABL above and the constraint that the profiles must match continuously at the IBL interface. This approach assumes that the dynamics above the IBL are unaffected by its formation. This approximation can be valid only fairly close to the transition region. More sophisticated model studies (e.g., Claussen,

1987) have shown that substantial mean vertical motions are also induced by the transition, even outside the IBL. Because the IBL exists for such a short distance in convective conditions, the formation and growth of the stable IBL are more critical. Here a key issue is the physics of the entrainment processes at the top of the IBL and the associated induced vertical velocity fields.

The Inhomogeneous Atmospheric Boundary Layer

Understanding ABL development and evolution in regions of abrupt or gradual changes of surface properties (coastal zones, ice-to-water surface transitions, ocean surface temperature fronts, etc.) involves consideration of horizontal advection, baroclinic forcing, nonequilibrium turbulence effects (for instance, the time derivative of the turbulent kinetic energy is not negligible), the special influence of local clouds, and fully three-dimensional dynamical processes.

A simple way to view this three-dimensional problem is to break it down into a conceptual model: a surface grid with each surface point occupied by a mean and turbulence profile, governed by one-dimensional turbulent mixing processes. Adjacent grid points are coupled in the normal manner through horizontal advection and horizontal pressure gradients. This approach has enjoyed some success with mixed layer models (Overland et al., 1983; Reynolds, 1984; Stage and Businger, 1980; Steyn and Oke, 1982) and higher-order closure models (Bennet and Hunkins, 1986; Tjernstrom, 1990; Wai and Stage, 1990). However, decoupling of the turbulence dynamics from the horizontal structure is a simplifying assumption that has never been tested. Clearly, similarity approaches (e.g., mixed layer models) that are tuned to quasi-equilibrium conditions are of limited applicability in inhomogeneous conditions. However, these limits have not been established. It may be that the techniques used in highly inhomogeneous urban boundary layer models (e.g., Uno et al., 1989) are adaptable to coastal problems.

Both land-sea breeze cycles and cold air outbreaks have been examined with models, but a comprehensive and extensive program to compare model results with measurements has not been attempted. Baroclinic effects associated with a sloping inversion (Brost et al., 1982; Overland et al., 1983) are known to be substantial, especially in west coast regimes, but they too are virtually unstudied.

The relative lack of experimental studies of boundary layer physics in coastal zones is not the only reason to question present-day boundary layer models. A recent LES study (Moeng and Wyngaard, 1989) of second-order closure parameterizations suggests that their rather modest successes in homogeneous conditions (e.g., Holt and Raman, 1988) are not expected to carry over to the coastal regime because homogeneous conditions do not severely test the parameterizations. To quote Moeng and Wyngaard (1989, italics added for emphasis), "Most observational and model studies show that *in the absence of abrupt changes in boundary conditions*, the heat flux profile is indeed essentially linear in the mixed layer. Thus, given the proper boundary conditions, second-order models will tend to have the correct vertical profile of buoyant production rates within the mixed layer, *regardless of the fidelity of their closure parameterizations*." Moeng and Wyngaard also point out that under homogeneous conditions the mean value of the rate of dissipation of turbulent kinetic energy in the mixed layer is also nearly independent of closure approximations.

The importance of local coherent wind circulations generated by heterogeneous surfaces also must be assessed. Existing work (e.g., André et al., 1990; Hadfield et al., 1991, 1992; Pielke et al., 1991; Walko et al., 1992) suggests that, if the spatial scale of the inhomogeneity is sufficiently large, a well-defined atmospheric circulation can develop. Development of such coherent circulations is also dependent on the large-scale wind speed, with stronger winds inhibiting their development. Claussen (1991) has introduced the concept of a blending height that could be applied when coherent wind circulations are likely to occur. This height is correlated with the horizontal scale of variability and the aerodynamic roughness of the landscape. When this height nears the height of the boundary layer, adjacent boundary layers are not significantly homogenized, and the resultant horizontal gradient in boundary layer heating can produce a well-defined local flow. When this height is much less than the planetary boundary layer height, however, the horizontal mixing of the boundary layer precludes coherent circulations. We need to understand the conditions under which horizontal inhomogeneities in surface heating and cooling generate coherent wind circulations.

Boundary Layer Clouds

Clouds within the ABL greatly complicate the physical processes because they represent a form of vertical and horizontal inhomogeneities, significantly affect the dynamics, and couple strongly with atmospheric radiation. For example, radiative heating effects of marine stratocumulus clouds can cause stress divergence to vary within the lower part of the ABL. During the day, if the cloud layer warms faster than the subcloud layer, the cloud may become decoupled, with a corresponding increase in wind stress and heat flux divergence between the top of the subcloud layer and the surface (e.g., Hignett, 1991; Rogers and Koracin, 1992).

Stratiform clouds are persistent features of cool upwelling coastal regions, such as the west coast of the United States, and cool climate regions such as the Arctic. Recent studies of the global radiation budget have highlighted the possible role of stratiform clouds as an ameliorating influence on the warming of the atmosphere by high-altitude clouds (Ramanathan et al., 1989). Although a number of experiments provide considerable insight into the processes that control the development and dissipation of stratiform clouds (Randall et al., 1984), the effect of variability of the coastal ocean, topography, and the marine ABL on these clouds has received much less attention. The effect of coastally trapped waves on the depth of the marine layer may also play an important role in the persistence of these clouds; the effect of the sea breeze on subsidence and the mesoscale pressure gradients may also be an important mesoscale process that controls the life cycle and fractional coverage of coastal stratus. In turn, the stratus clouds can substantially modulate the sea breeze cycle by moving ashore and reducing inland solar-induced convection. Skupniewicz et al. (1991) have presented the only measurement and model examination of the diurnal evolution of the coastal stratocumulus cloud edge and its dramatic effect on sea breeze front dynamics.

SUMMARY AND CONCLUSIONS

Understanding of coastal boundary layer processes will be improved by general advances in boundary layer science. However, some problems specific to the coastal regime require immediate study. Theoretical developments, modeling studies, and field measurement programs are required to explain key unknown properties and processes involving the ABL. The panel recommends the following:

• Studies should be conducted to determine the properties for inhomogeneous and nonequilibrium conditions, including suitable surface flux and mixed layer similarity parameterizations, and the general relationships of the ensemble average first- and higher-order turbulence variables.

• Further research on the fundamental relationships among the ocean wave spectrum, surface fluxes, and bulk ABL properties should be conducted.

• Studies should be carried out to determine the physical process of the growth of the top of the stable IBL, including entrainment and induced mean vertical velocity effects, and the decay processes of turbulence above a newly formed IBL.

• Investigations to clarify coastal marine stratocumulus and overland fair weather cumulus cloud regimes and their influence on land-sea breeze cycles should be conducted.

• Studies should be undertaken to determine the spatial scale at which horizontal inhomogeneities in surface heating and cooling become large enough to generate coherent wind circulations.

3

Thermally Driven Effects

Differences in land and sea surface temperature and heat flux result in direct, thermally driven wind systems over a spectrum of temporal and spatial scales. The best known among these are the mesoscale land and sea (lake) breeze circulation systems (see, e.g., Defant, 1950), which are inherently diurnal in nature. Much less studied are the regimes induced by warm ocean waters adjacent to a large cold land mass. These circulations, which are not diurnal in nature, are termed coastal fronts. Even less well understood are systems present over the coastal ice-land boundary. As an example, in arctic regions, offshore katabatic winds are believed to play a key role in forming and altering polynyas and leads in coastal ice sheets.

THE LAND BREEZE AND THE SEA BREEZE

The land-sea breeze system (LSBS) typifies the class of mesoscale atmospheric systems induced by spatial inhomogeneities of surface heat flux into the boundary layer. The LSBS has been identified since the time of the classical Greeks (circa 350 B.C.). By the late 1960s, identifiable literature references exceeded 500 (Baralt and Brown, 1965; Jehn, 1973). Of all mesoscale phenomena, the LSBS over flat terrain has been among the most studied observationally, analytically, and numerically. This is undoubtedly a result of their geographically fixed nature, their frequent occurrence, their ease of recognition from conventional observations, the concentration of observers in coastal zones, and their importance to local weather and climate.

By way of definition for this review, a land and sea breeze is a diurnal thermally driven circulation in which a definite surface convergence zone exists between air streams having over-water versus overland histories. These breezes are differentiated from the sustained onshore-offshore winds driven by the synoptic pressure field, which are termed sea-land winds. During lighter synoptic wind regimes, perturbations induced by the coastal discontinuity are often detectable but may not always result in a coherent recirculating wind system. The effects of the LSBS are many, including significantly altering the direction and speed of the ABL winds; influencing low-level stratiform and cumuliform clouds; initiating, suppressing, and modifying precipitating convective storms; recirculating and trapping pollutants released in or becoming entrained into the circulation; perturbing regional mixing depths; and creating strong near-shore temperature, moisture, and refractive index gradients. Improved understanding of LSBS should enhance applications to a wide variety of commercial, industrial, and defense activities (Raman, 1982).

Sea-lake breeze inflow layers can vary from 100 m to over 1000 m in depth. Inland frontal penetration can vary from less than 1 km to over 100 km, with propagation speeds ranging from nearly stationary to >5 m/sec. The offshore extent of the inflow layer is less well known. Peak wind speeds are typically less than 10 m/sec. The overlying return flow layer depth is generally twice that of the inflow, but is often difficult to differentiate from the synoptic flow. Given the difficulty of measuring atmospheric mesoscale vertical motions, little is known about the detailed structure of updrafts associated with the sea breeze front. Some observational evidence from gliders, tetroons, Doppler lidar, etc., has suggested organized frontal zone motions of several m/sec. The coarse mesh size (often >5 km) used in most mesoscale simulations tends to portray peak vertical motions in the tens of centimeters per second range. More recent modeling studies (Lyons et al., 1991a, b) suggest that the sea breeze convergence zone is at times extremely narrow (perhaps <1000 m) and may produce regions of organized mesoscale ascent ranging from 1 to 4 m/sec. Even less is known about the broader and weaker subsidence regions offshore. Needed are improved measurements of vertical motions associated with the LSBS and companion modeling studies in which the mesh sizes used are adequate to resolve the observed features (Figure 3.1). Recent extremely fine-mesh two-dimensional simulations by Sha et al. (1991) resolved complex Kelvin-Helmholtz instabilities and other characteristics resembling a laboratory gravity current (Simpson, 1982).

Circulations over large lakes are very similar to their oceanic counterparts. Smaller lakes, estuaries, and larger rivers also can significantly perturb the regional flow. The interactions between synoptic flow and water body size and orientation require additional study. Recent numerical simulations have suggested that surface heat flux differences of 100 W/m^2 or more over several tens of kilometers can generate sea-breeze-like circulations (Segal et al., 1988). These physiographic mesoscale circulations can result from differences in soil type, land use, soil moisture (from irrigation and precipitation), snow cover, smoke/haze, and cloudiness. Additional field programs investigating mesoscale processes in both LSBS and physiographic systems are warranted as a test of the validity of available models.

The LSBSs are not exclusively "clear weather" phenomena; they substantially affect and interact with various cloud types. The modifications of fog and stratus in the LSBSs over the Great Lakes, Gulf, and Atlantic coasts have received less attention than those along the Pacific coastline. Advection through the coastal domain of middle- and high-level cloud decks, large smoke plumes, and urban- and regional-scale photochemical and sulfate hazes can materially affect the energetics of the LSBS, although these impacts have not been quantified. The LSBS profoundly affects the formation and fate of shallow convective clouds. Cumulus suppression in subsiding regions of the sea breeze cell has long been noted in satellite imagery. Under very light wind conditions, cumulus growth is enhanced within the sea breeze frontal zone updrafts. When the prevailing regional flow advects small convective clouds seaward across the frontal zone, the responses are more complex. Dissipation often occurs, but the relative roles played by subsidence versus disruption of cloud-capped thermals rooted in the surface superadiabatic layer are uncertain. Studies of the interactions of large eddies or convective thermals with the sea breeze frontal zone are warranted.

The impact of the sea breeze front on deeper (precipitating) convective clouds is more complex. At midlatitudes the sea breeze can either enhance or weaken convective storms (Chandik and Lyons, 1971). On a scale of tens of kilometers, the Florida sea breeze has been found to trigger the general development of deep convection (Burpee and Lahiff, 1984). Thunderstorm development is intermittent along such frontal zones, and it is uncertain whether perturbations in the frontal convergence zone or localized responses to inhomogeneities in the surface energy budgets (or both) cause individual storms to form. Fine-mesh numerical modeling studies suggest that different spatial scales of topography and shoreline geometry produce a spectrum of convective-scale responses. Sea breeze thunderstorms affecting the Kennedy Space Center (Lyons et al., 1992) are initiated by both the east and west coast sea breezes. In addition, however, strong local convergence onto features such as Merritt Island trigger smaller-scale thunderstorms embedded within the larger sea breeze circulation (Figure 3.2). The complex feedbacks between the precipitating clouds and the LSBS are only partially understood. Also, convective responses to sea breezes over mountainous islands such as those found in the "marine continent" of southeast Asia require further study. Sea breeze thunderstorms contribute approxi-

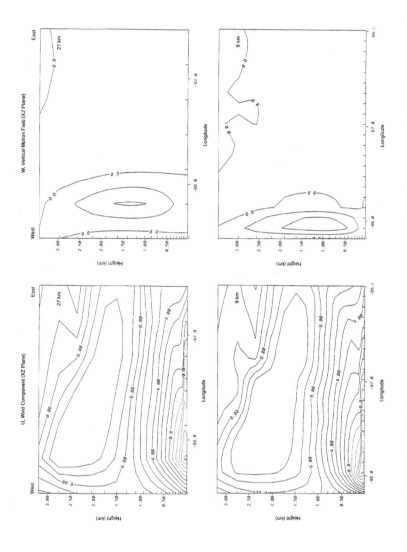

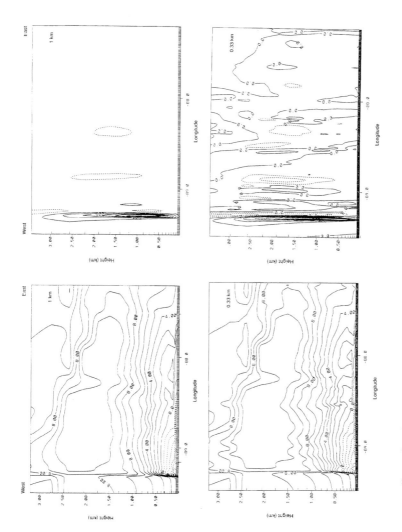

FIGURE 3.1   Four two-dimensional mesoscale model realizations of a lake breeze in a 3000-meter deep east-west plane across southern Lake Michigan using 27-, 9-, 1-, and 0.33-km horizontal mesh sizes. All frames are at 1500 LT. Shown are U, the wind component, 1 m/sec isotachs (left); and W, the vertical motion field (right). Peak W values increase from 23 cm/sec (at 27 km mesh size) to 212 cm/sec (at 0.33 km mesh size). Negative values are shown as dashed lines.

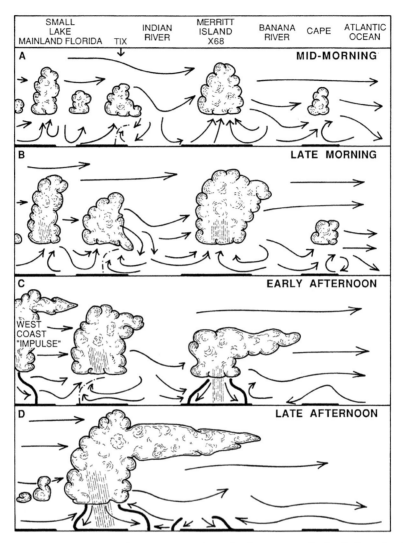

FIGURE 3.2 Response of deep convective clouds to coastal circulations in the vicinity of the Kennedy Space Center. In addition to the primary east and west coast sea breezes formed by the contrast between the Florida peninsula and surrounding ocean, numerous lakes, surface land use inhomogeneities, estuaries, and islands perturb the mesoscale flow. The convective response is more complex than suggested by earlier sea breeze thunderstorm studies. As an example, while the general Atlantic sea breeze (ASB) develops, enhanced convergence onto Merritt Island triggers rapid growth of a small thundershower by late morning. Widespread convection along the ASB does not develop until the late-afternoon approach of an impulse associated with the west coast sea breeze (graphics courtesy of Cecil S. Keen).

mately 40 percent of Florida's rainfall and probably more in many other coastal regions. Sea breeze convective storm development is sensitive to small variations in middle-tropospheric temperature lapse rates and moisture, such as those postulated in global greenhouse warming scenarios. Thus, middle tropospheric changes that affect this significant source of precipitation in tropical coastal areas is potentially important.

Except for a few recent studies (Ohara et al., 1989), the land breeze is understood even less. Wind speeds are typically <5 m/sec, and the offshore flowing layer is often <100 m deep. The land breeze can often be commingled with stronger and deeper terrain-induced katabatic flows. The formation of thunderstorms associated with the offshore boundary of the land breeze over Gulf and Atlantic coastal waters has been addressed only in very cursory ways. Intense Great Lakes snow squalls also interact in complex ways with the land breeze (Passarelli and Braham, 1981).

The transition of the land breeze into a sea breeze, occurring offshore, is largely undocumented. The breakdown of the land-lake breeze front is not well understood. Sometimes it retreats offshore as a distinct front, while at other times it simply pushes inland and dissipates after sunset. On other occasions, strong onshore flow may continue over coastal regions until past midnight local time (the "fossil" sea breeze). Comprehensive studies of the LSBS through consecutive diurnal cycles are desired, with emphasis on the land breeze and the morning and evening transition periods.

The LSBS and many other similar mesoscale circulations are poorly resolved in conventional weather-observing network systems, creating serious problems in operational forecasting. Local forecasters employ simple forecasting techniques using synoptic observational data (Lyons, 1972) to predict potential sea breeze occurrences. The character of the LSBS is controlled by a variety of factors, including land-sea surface temperature differences; latitude and day of the year; the synoptic wind and its orientation to the shoreline; the thermal stability of the lowest 200 to 300 mb of the atmosphere; patterns of land use and soil moisture; surface solar radiation as affected by haze, smoke, stratiform, and convective cloudiness; and the geometry of the shoreline and complexity of the surrounding terrain. Many of these factors are considered within mesoscale numerical modeling systems that are well suited to land-sea breeze simulation.

Our understanding of the LSBS is not comprehensive, being largely confined to idealized conditions. When large-scale winds are virtually nonexistent over an infinitely long, two-dimensional flat coastline, it is comparatively easy to describe the basic dynamics of the LSBS (Defant, 1951). Numerous analytical studies of sea breeze phenomena have been conducted (see, for example, Haurwitz, 1947, and Rotunno, 1983). Nonlinear numerical modeling studies using two-dimensional models have been summarized by Pielke (1984). Newer nonhydrostatic, fine-mesh, nested-grid numerical models will be applied profitably to further studies of these types. Many previous modeling efforts emphasized coastal circulations in which the forcing functions were spatially homogeneous and temporally steady-state or diurnally varying. Observational and modeling studies need to be extended to coastal circulations occurring with nonhomogeneous and nonstationary synoptic environments, irregular shorelines and complex topography, and heterogeneous land use or land characteristics and soil moisture. Additional modeling challenges include accounting for the advection of middle- and upper-level clouds through the domain; changes in soil moisture; dynamic feedback between the LSBS and deep convective storms; and turbidity due to regional smoke, pollution, fog, and haze. There is evidence that gravity waves are excited by regional air flowing over the sea breeze, which is dynamically equivalent to a mountain. These features are worthy of continued investigation.

While hemispheric and synoptic-scale numerical forecasting became well established in the 1950s, it was not until the mid-1980s that more powerful computers allowed organized attempts at operational mesoscale forecasting of coastal circulations and their effects (Lyons et al., 1987). Affordable high-speed computing and increasingly sophisticated numerical modeling techniques now allow extended experiments in operational coastal zone wind forecasting to be undertaken, yielding excellent opportunities to test the breadth and depth of our understanding of the LSBS.

Interaction of the LSBS with the urban heat island and greatly enhanced roughness lengths in large cities has been studied in New York, Tokyo, Toronto, and elsewhere. Studies of interacting sea breeze and topographically forced flows (such as the Catalina eddy) are yielding improved understanding of the complex interactions between mesoscale systems. Even with the large number of studies in coastal Southern California, the development and morphology of sea and land breeze circulations in mountainous coastal terrain warrant much additional attention.

Coastal thermally driven mesoscale circulations interact with smaller-scale surface-atmosphere energy exchange processes, cloud systems on a variety of scales, and the larger-scale synoptic patterns in which the mesoscale circulations are embedded. The LSBS represents a challenging problem for future observational and modeling programs, since it embodies many of the complex issues involved in atmospheric-scale interactions.

COASTAL FRONTS

When air over land is colder than air over the sea for extended periods, the direct circulation that develops across a coast does not normally exhibit typical diurnal characteristics. A longer-lived cousin of the land and sea breeze front, called a coastal front (Bosart et al., 1972), can form and remain quasi-stationary, parallel to the coast, for several days. Coastal fronts have been reported and studied in many parts of the world, including the East and Gulf coasts of the United States (Bosart, 1975, 1981, 1984), the Black Sea (Draghici, 1984), Norway (Økland, 1990), the Netherlands (Roeloffzen et al., 1986), and Japan (Fujibe, 1990). The importance of coastal fronts for freezing rain and coastal cyclogenesis is discussed in Chapter 5.

Along the Carolinas, coastal fronts are typically 1000 km long, with temperature contrasts as large as 20°C. Carolina coastal fronts tend to form within the boundary layer temperature gradient produced by differential heating of air across the margin of the Gulf Stream. This process has been studied by Riordan (1990), who used radar, ship, buoy, and aircraft data from the Genesis of Atlantic Lows Experiment (GALE). Frontal formation was found to be a discontinuous process, with the front forming in segments aligned with bands of shallow convection (Figure 3.2). The inland motion of the front was also discontinuous, for reasons that are not understood. This behavior contrasts with that of New England coastal fronts, which have been found to form along the coast and retain their identity as they move inland (Nielsen, 1989).

A wide variety of triggering mechanisms have been shown to provide the sustained differential heating or confluence necessary for coastal fronto-genesis. The most common occurs when a cold anticyclone approaches a coastline and winds become parallel to the coast. Air over land remains cold, while air just offshore continuously receives large heat fluxes from the sea surface. For example, over the Gulf Stream convective rainbands often form which may be accompanied by considerable lightning activity (Hobbs, 1987; Biswas and Hobbs, 1990). Other processes that are favorable to coastal frontogenesis are frictional retardation and turning of the wind, upstream blocking of cold air, or lee convergence.

ICE-EDGE BOUNDARIES

An understanding of the meteorology of coastal regions where sea ice is present is important for navigation, exploring for mineral resources, coastal biological activity, and modeling sea ice and climate. In the present discussion, only ice-land boundary regions will be considered. This includes almost all the Antarctic ice cover and the ice cover overlying the high-latitude continental shelves in the northern hemisphere. While many characteristics of the marginal ice zone (the boundary between sea ice and open ocean; see Johannessen et al., 1988) are similar to those of coastal ice-edge boundaries, the marginal ice zone will not be addressed here.

A summary of meteorological processes occurring at ice-edge boundaries is given by Barry (1986). Barry concluded that our basic knowledge of meteorological conditions over ice-edge boundaries is very limited. The recent Marginal Ice Zone Experiment (MIZEX) off the East Greenland and Bering seas has provided insight into processes occurring at ice-ocean boundaries, but no such program has been undertaken for land-ice boundaries. With the exception of Antarctic katabatic winds (see Chapter 4), relatively little research has been conducted on mesoscale and small-scale coastal processes at the land-ice boundary. Many of the same physical processes and phenomena, which are described elsewhere in this report, such as land and sea breezes, occur in ice-edge coastal regions. However, there are unique processes occurring at the ice edge, particularly during the cold seasons of the year, that are associated with the characteristics of sea ice. A summary of the observed features of coastal sea ice is provided by Wadhams (1986) in the context of the seasonal sea ice zone.

Even during winter, regions of open water occur between sea ice around coasts. An overview is given by Smith et al. (1990b) of polynyas and leads, which are openings in pack ice due to ice drift divergence and local melting. Offshore katabatic winds (see Chapter 4) are believed to play an important role in the formation and maintenance of polynyas and leads. Particularly during winter, leads and polynyas are a major source of exchange of heat, moisture, and gases between the ocean and atmosphere. Polynyas and leads are sites of active brine formation, affecting the local water density structure and current field and cumulatively affecting the structure of the halocline. Leads and polynyas serve as corridors for migration of marine mammals. During spring localized plankton blooms occur, which are important biologically and are also possibly important as a local source of cloud condensation nuclei (see Chapter 6).

A feature that occurs along the Antarctic coast is the presence of ice shelves, over which glacier ice streams into the sea. The largest ice shelves in the Antarctic are the Ross and Ronne-Filchner. Ice shelves determine the capability of the fast-flowing internal ice streams associated with marine ice sheets to disperse the glacier ice rapidly into the surrounding ocean. Marine ice sheets are characterized by being grounded on beds well below sea level. If the backstress exerted on the ice stream by the ice shelf is insufficient, accelerated discharge of land ice through ice streams to the sea may result in the collapse of the marine ice sheet (Binschadler, 1991).

Leads and polynyas affect the atmosphere in the ice-edge coastal regions in the following ways. Extreme sea-air temperature differences (20° to 40°C) are commonly associated with leads and polynyas during winter, and heat fluxes of several hundred watts per square meter are typical (Smith et al., 1990b). The sensible heat flux in air is several times larger than the latent heat flux because of the relatively low value of saturation-specific humidity at the freezing point. Schnell et al. (1989) found that wide leads and polynyas release enough energy to create buoyant plumes that penetrate the boundary layer; in one case a hydrometer plume reached a height of 4 km. Aircraft lidar measurements indicated the presence of small ice crystals in the plume, which are believed to modify substantially the local radiative balance. In particular, the larger leads and polynyas are likely to provide substantial amounts of water vapor, especially to the wintertime polar atmosphere, contributing to the presence of widespread low-level stratus clouds in these regions. Convection, associated with substantial precipitation, from larger polynyas seems likely; this precipitation has the potential to significantly influence the accumulation of snow both on glaciers and on the sea ice itself. The panel notes that the forthcoming Lead Experiment (LEADEX) in the Beaufort Sea, although not occurring in the coastal zone, will address some of these issues and improve our general knowledge about leads. At the same time, the state of the sea ice, including the presence of leads and polynyas, is strongly dependent on atmospheric processes (see, e.g., Hibler, 1979). The atmosphere influences the state of the sea ice both thermodynamically (e.g., via radiative heat, sensible heat, and latent heat fluxes) and dynamically (e.g., via surface wind stress).

SUMMARY AND CONCLUSIONS

Some existing gaps in scientific understanding associated with thermally driven effects may be addressed through modeling studies and field programs. To spur progress we recommend the following:

• Observational and modeling studies of the LSBS should be extended to cover the entire diurnal cycle, with emphasis on improving knowledge of offshore regions, the morphology and dynamics of the land breeze, and the formation and breakdown of the sea breeze front.

• Remote sensing techniques and fine-mesh mesoscale numerical models should be applied to better understand the finer-scale, three-dimensional structure of the sea breeze front, its associated mesoscale vertical motions, and the development of internal boundary layers above complex coastlines and heterogeneous surfaces.

• Research should be directed to understand the three-dimensional LSBS interactions with inhomogeneous and time-dependent synoptic flows, non-uniform land and water surfaces, irregular coastlines, and complex terrain, as well as the dynamic feedbacks between the LSBS and stratiform clouds and precipitating and nonprecipitating convective cloud systems.

• The geographical distribution of coastal front occurrences, their spatial coverage, and their modes of propagation should be documented and their variability assessed.

• A combination of case studies and model simulations should be conducted to determine the site-specific, large-scale conditions leading to coastal front formation, which is often difficult to observe directly in real time because coastal fronts tend to form offshore and sometimes remain nearly stationary.

• Studies should be supported to elucidate processes of heat and moisture fluxes into the atmosphere from leads and polynyas, particularly in the presence of extreme horizontal thermal discontinuity.

• Interactions between the atmosphere and sea ice on the mesoscale in the coastal zone should be examined.

4

The Influence of Orography

INTRODUCTION AND BASIC PARAMETERS

The transition from a nearly flat ocean to land in the coastal zone is often accompanied by major changes in elevation. Flow over and around such changes in orography in a rotating stratified fluid represents one of the classic problems in meteorology and oceanography. For a far-field wind perpendicular to a barrier, the horizontal extent and magnitude of upstream modification of the flow pattern in response to the barrier should be determined. There is also a downwind modification of the flow, which is not, in general, symmetric with the upwind influence.

Low-level air flow is generally blocked by a mountain when the parameter known as the Froude number is less than unity (Smith, 1979, 1989). The Froude number provides a measure of the relative importance of potential and kinetic energy in flow around and over obstacles. It is defined by:

$$F_r = U/h_m N, \tag{1}$$

where N is the static stability and is equal to $(-g\partial_z \theta/\theta_o)^{1/2}$, U is the speed of the free air stream, h_m is the height of the ridge, θ_o is the constant mean potential temperature, and g is gravity. For typical atmospheric stratification of $N \cong 10^{-1}$ to 10^{-2}, an elevation of only 100 m is often sufficient to cause "blocking" of the onshore flow at low levels. Such blocking occurs along the west and east coasts of the United States; along the east coast, it is often referred to as cold air damming. Leeside effects can be important for flow directed offshore; for island wakes; in regions where the coastline curves, such as Southern California; or where there are inland marine regions, such as Puget Sound, San Francisco Bay, and southeast Alaska.

In the coastal zone it is not necessary to have an upstream velocity directed toward the mountains for the orography to influence coastal winds. If a localized region of high or low pressure is generated in the coastal zone, it will, under certain conditions, be trapped and propagate along the coastline within the coastal zone. This is a common phenomenon, for example, along the coasts of California and Australia.

The Froude number considers the relative importance of vertical displacement of isentropic surfaces in flow around and over obstacles. A second factor is the influence of the earth's rotation on upstream flow deceleration (Queney, 1948). One can consider the influence of rotation through a Rossby number:

$$R_m = U/fl_m, \tag{2}$$

where U is the upstream velocity, f the Coriolis force, and l_m is the half-width of the ridge; little flow deceleration is found when R_m is less than unity. Numerical simulations by Pierrehumbert and Wyman (1985) and trajectory analyses by Chen and Smith (1987) suggest that in the region of steep topography the deceleration zone will grow upstream to a width of:

$$l_R = Nh_m/f. \tag{3}$$

This parameter l_R is known as the radius of deformation. Steep topography is defined by the nondimensional slope, $(h_m/l_m)(N/f)$, being greater than 1. For the coastal case, l_R is often on the order of 50 to 150 km and $l_R > l_m$; this contrasts with broad mountain ranges such as the Rockies with l_m on the order of 500 km. In the broad mountain case, $l_m > l_R$, the flow stays quasi-geostrophic with $R_m < 1$ (i.e., wind blows perpendicular to the pressure gradient as it flows over the topography, with little upstream influence). The coastal region, however, is often in the knife-edge mountain case, $l_R > l_m$, where $R_m > 1$. Here one expects the coastal mountains to represent a wall, and the momentum balance in the along-shore direction near the wall is not expected to be geostrophic. The smoothed topographies in current-generation numerical weather prediction (NWP) models do not even qualitatively represent knife-edge slopes and thus do not correctly include coastal phenomena.

To further delineate the influence of orography on coastal meteorology, let L be the scale for motion in the along-coast (y) direction, and l be the scale in the cross-shore direction ($-x$), where

$$l_m < l < L. \tag{4}$$

We can nondimensionalize the equations of motion in the following manner. The cross-shore wind component, u, and along-shore wind component, v, are scaled by UL/l, time by l/U, vertical distances by $D = fl/N$, and pressure p by $P_ofl U$. The equations of motion for such a shallow system become (Overland, 1984):

$$R_l\left(\frac{l}{L}\right)^2\left[\frac{du}{dt}+C_D'u\right]=v-\frac{\partial P_o}{\partial x}-\frac{\partial p}{\partial x} \quad \text{(cross-shore)} \qquad (5)$$

and

$$R_l\left[\frac{dv}{dt}+C_D'u\right]=-u-\frac{\partial P_o}{\partial y}-\frac{\partial p}{\partial y} \quad \text{(along-shore)}, \qquad (6)$$

where $R_l = U/fl = V/fL$ is a coastal Rossby number, $C_D' = C_D(U + V)/D$ is a coastal drag coefficient that indicates the relative importance of surface friction, and P_o is a synoptic-scale pressure imposed on the coastal layer. The term $\partial P_o/\partial y$ on the right-hand side of Eq. (6) nondimensionally equals unity and represents the along-shore pressure gradient associated with the incident geostrophic wind, U. For many coastal problems, the left-hand side of Eq. (5) is small, even though R_l is on the order of Eq. (2); the flow in the along-shore direction is in geostrophic balance. The left-hand side of Eq. (6), however, is on the order of Eq. (2), and v can exhibit accelerations in response to the imposed along-shore pressure gradient. Small l/L and $R_l \sim 1$ are the heart of the coastal zone semigeostrophic approximation. The only remaining free parameter is the nondimensional mountain height

$$h_m/D = (N/f)(h_m/l). \qquad (7)$$

This may also be written as R_l/F_r. Thus, the coastal mountain problem can be specified in terms of R_l and F_r, a Rossby number and a Froude number. Note that for $h_m/D = 1$ (i.e., steep topography), the offshore length scale is defined by $l = l_R$, the Rossby radius of deformation, which scales coastal influence offshore of mountainous coasts to be 10 to 100 km.

The foregoing discussion suggests that mesoscale meteorological features (10 to 100 km) are generated in the vicinity of coastal orography, and that ageostrophic motions are anticipated in the along-shore direction. However, the upstream flow is seldom stationary and uniform; vertical stratification is seldom constant. While theoretical considerations define the scales and processes important to the coastal zone, they are less successful in fully explaining particular case studies (Walter and Overland, 1982; Mass and Ferber, 1990).

LOW FROUDE NUMBER FLOW: TRAPPED PHENOMENA

Isolated Response: Kelvin Wave and Gravity Current

Along subtropical mountainous coastlines such as California and Australia, subsidence in the subtropical high pressure often develops a strong marine inversion structure below the height of the coastal topography. This case is the most studied of orographic coastal phenomena. In this case the stability scale can be replaced by the difference in temperature, $\theta_2 - \theta_1$, across the inversion occurring at height h_i. The coastal zone semigeo-strophic equations, (5) and (6), admit Kelvin wave solutions

$$h_i'/D = e^x\, G(y - t),\tag{8}$$

where G is an arbitrary function. The solution is trapped to a unit distance from the coast (i.e., a Rossby radius) and propagates at a unit speed that in dimensional terms has the phase speed $c = fl_R$. On the other hand, if the equations are initialized with a density front, a nonlinear gravity current, with some Kelvin-like aspects, can form. Along-shore disturbances can also be forced by an along-shore pressure gradient $\partial P_o/\partial y$ imposed by the synoptic-scale pressure field above the coastal layer. The existence of this class of mesoscale coastal features has been documented in California (Beardsley et al., 1987; Dorman, 1985, 1987; Mass and Albright, 1987; Winant et al., 1988; Zemba and Friehe, 1987) and Australia (Holland and Leslie, 1986). Figure 4.1 shows a typical summer trapped feature along the west coast of the United States. The GOES visible imagery (Figure 4.1a) shows a wedge-like feature that propagates northward from central California to British Columbia in about 2 days. Note that in the sea-level pressure analysis (Figure 4.1b) the wind shifts from northerly to southerly with the passage of

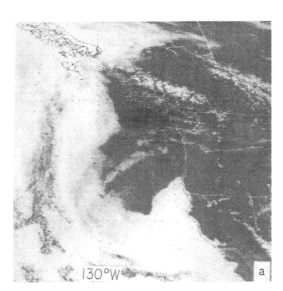
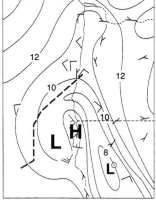

FIGURE 4.1 Typical summertime northward surge of marine air trapped to the west coast: (a) GOES visible image 1800 GMT, May 27, 1983; (b) observed wind and subjective sea-level pressure analysis for the same time based on satellite images and available surface synoptic data. Note that there is wind shift in the coastal zone as the surge propagates northward (after Mass et al., 1986).

the feature. The case for South Africa (Bannon, 1981; Gill, 1977; Reason and Jury, 1990) is quite distinct from the Australia and California cases (Reason and Steyn, 1990) (see Table 4.1). Calculation of the Rossby number based on half the mountain width yields 0.1, so dynamics lie within the quasi-geostrophic regime (i.e., small mountain slope) and blocking of an incipient flow will not persist (Pierrehumbert and Wyman, 1985).

While these isolated wave/frontal features are perhaps the most obvious of coastal phenomena, understanding their source mechanisms and composite nature (density flow versus propagating wave) is uncertain in any particular realization. These isolated trapped phenomena are generally initiated by changes in the synoptic-scale flow. The climatology of such changes is not well documented and is an area for further research. There is also a need to understand all factors that contribute to the depth and spatial variability of coastal marine stratus and fog as a result of interactions among wave dynamics, radiation, and cloud processes.

Damming

The case of damming refers to blocked winds on the windward side of a mountain for uniform onshore flow or modification of a frontal feature by coastal orography. This phenomenon is less well documented for coastal regions than for other mountain regions. The balance of the wind and pressure (mass) fields within the storm is disrupted at the coast. As a result, the path of the storm can change abruptly, and, in certain instances, barrier jets and enhanced surface winds can develop in the coastal zone. Mass and Ferber (1990) show the development of ridging along the coast of western Washington state with the approach of a cold frontal system (Figure 4.2). When this orographically induced pressure field is added to the synoptic-scale pressure field, it produces large along-shore pressure gradients, which the momentum field responds to by producing an along-shore wind jet that is stronger than the winds in the weather system farther offshore. Similar super-geostrophic winds have been observed at coastal stations along Alaska (Businger and Walter, 1988; Reynolds, 1983). What is not known for these cases is how the storm system itself is modified by the presence of a coastline and orography. This interaction of storms and orography to produce coastal jets is a major area for research.

A related phenomenon along the east coast of the United States is Appalachian cold air damming (Bell and Bosart, 1988; Xu, 1990). These episodes arise when there is high pressure over New England and onshore flow toward the Appalachians with an estimated Froude number of 0.3 to 0.4. A semigeostrophic system is set up on the eastern slopes with a low-level wind maximum parallel to the ridge. This wind maximum is fed by a pool of cold air from the north, which creates a cold dome along the eastern slopes. In the Bell and Bosart (1988) study, over-water winds were not

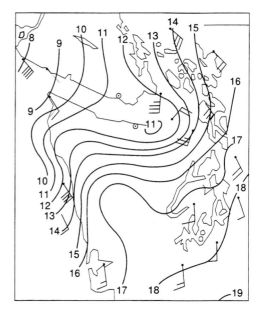

FIGURE 4.2 Sea-level pressure analysis and surface winds for Washington state, 2100 GMT, March 5, 1988. Note the coastal wind jet, which developed as an ageostrophic response to the increased along-shore pressure gradient (after Mass and Ferber, 1991).

directly considered; however, it is clear from their analyses that the cold air damming region extended to coastal weather stations.

Results from research on inland mountain systems may be relevant to damming, particularly the Alpine experiment (Chen and Smith, 1987; Davies and Pichler, 1990; Schuman, 1987, for example), which considers blocking and synoptic weather system/orographic interaction in the Alps. Another case is along the Sierra Nevada Mountains (Parish, 1982). One difficulty is that these cases may have more gentle slopes than those at coastlines and thus more of a quasi-geostrophic than semigeostrophic response. A summary of R_m and F_r for several cases is presented in Table 4.1. Poor verification of coastal weather forecasts is often attributed to the formation of mesoscale systems by the interaction of storms with coastal orography and to the feedback of these features on storm intensity within the coastal zone (± 100 km), yet even basic documentation of this interaction and feedback is lacking (Bane et al., 1990).

Gap Winds

A special case of trapped phenomena is a sea-level channel between two mountainous coastlines where the width of the strait is on the order of the Rossby radius or less. This establishes a semigeostrophic system in the strait with winds accelerating down the strait in response to the synoptic-scale along-strait pressure gradient (Overland and Walter, 1981) (Figure 4.3). In a case

TABLE 4.1   Rossby (R_m) and Froude (F_r) Numbers, Coastal Mountain Half-Width l_m, Inversion Thickness h_i, and Reduced Gravity g' for Coastally Trapped Disturbances in Southern Africa, Southeastern Australia, California, and the Alps (noncoastal)

	Southern Africa	Southeastern Australia	California	Alps
R_m	0.1	0.8	1.3	2.0
F_r	0.33	0.18	0.21	0.4
R_m/F_r	0.30	4.44	6.19	5
l_m (km)	600	75	50	50
h_m (m)	450	800	700	2500
g' (m/s^2)	0.11	0.60	0.41	N = 0.01[a]

[a]N is the Brunt-Vaiasala frequency.

SOURCE: After Reason and Steyn (1990) and Pierrehumbert and Wyman (1985).

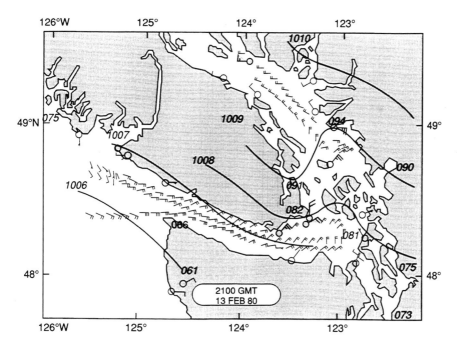

FIGURE 4.3   Sea-level pressure analysis for a gap wind event in the Strait of Juan de Fuca.   Flow accelerates along the strait, and there is an abrupt transition in wind speed and direction beyond the exit of the strait.   Small wind arrows are from aircraft-derived estimates; large wind arrows are coastal stations (after Overland and Walter, 1981).

study of an event in Shelikof Strait (Lackmann and Overland, 1989), it was entrainment of slower-velocity air from above the strait that provided the principal retarding effect opposing the along-strait pressure gradient. A feature of both these studies is a geostrophic adjustment process that occurs seaward of the exit to both straits, with front-like characteristics, rather than a smooth transition to the flow along the open coast.

MODERATE FROUDE NUMBER FLOW

For inland sounds and offshore flow, there can be a mesoscale response in the lee of coastal mountains (Dempsey and Rotunno, 1988). In general, this response is not symmetric with the windward response. Reed (1980), Smith (1981), Mass and Ferber (1990), Ferber and Mass (1990), and Walter and Overland (1982) discuss the response of flow in the lee of the Olympic Mountains in Washington (see Figure 4.2). Only a couple of times a winter will the incident flow (Froude number) be great enough to create an orographically induced mesoscale low-pressure center of significant amplitude in the lee of the mountains. During such an event, winds in Puget Sound will accelerate ageostrophically toward mesoscale low pressure, giving localized marine winds of 20 to 40 m/sec.

The Catalina eddy feature occurs south of where the California coastline changes from a north-south to an east-west orientation (Figure 4.4). A consensus (Bosart, 1983; Mass, 1989) is that under northerly flow the Santa Ynez Mountains north of Santa Barbara favor a lee eddy feature. Similar to the Puget Sound case, the induced mesoscale low creates an along-shore pressure gradient that can initiate northward propagation of trapped phenomena from the south (Dorman, 1987) and modify the position and magnitude of the lee eddy. However, a lee eddy is not a prerequisite, according to Mass (1989), for the formation of a Catalina eddy. An alternate explanation (Clark and Dembek, 1991) is that the eddies form just south of Santa Barbara and move southward. Sea breeze circulation often masks the structure of the Catalina eddy during the daytime.

Upwind and lee-side effects also influence coastal regions of islands and peninsulas. Nickerson and Dias (1981) and Smolarkiewicz et al. (1988) investigated the downwind regime of the Hawaiian Islands for very small Froude numbers (0.1 to 0.5). A pair of vertically oriented vortices form on the lee side of the island of Hawaii. These vortices form as a result of the tilting of horizontally oriented vorticity, produced baroclinically as isentropes deform in response to flow around the obstacle, rather than as a consequence of viscous boundary layer separation (Smolarkiewicz and Rotunno, 1989). Gaps in mountain chains can produce strong jets when there is a major cold reservoir on one side of the mountains (Macklin, 1988). Macklin et al. (1990) discuss one such jet that forms across the Alaskan

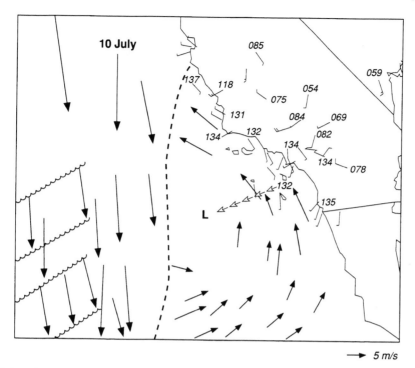

FIGURE 4.4 Cloud track wind estimates based on GOES visible imagery for a Catalina eddy case, July 10, 1987. L marks the position of the circulation center, with arrows showing the migration of the center from its formation location near Santa Barbara (after Clark and Dembek, 1991).

peninsula and continues across lower Cook Inlet with winds >20 m/sec. Bond and Macklin (1992) noted that, when the Froude number was greater than 1 for northerly winds incident on the Alaskan peninsula, there were mountain waves above the boundary layer and strong winds up to 34 m/sec along the south side of the peninsula. In a low Froude number case, strong winds were confined to regions downwind of low-level passes. Such lee-eddy features should be amenable to numerical modeling given sufficient vertical and horizontal resolution; models can include simplified cloud processes.

KATABATIC AND OTHER LOCAL WINDS

So far we have discussed phenomena with spatial scales of the order of the Rossby radius (50 to 150 km). However, there are smaller inlets and mountain passes, often perpendicular to the coastline, that can produce locally strong marine winds. If there is a cold reservoir of air inland, this air

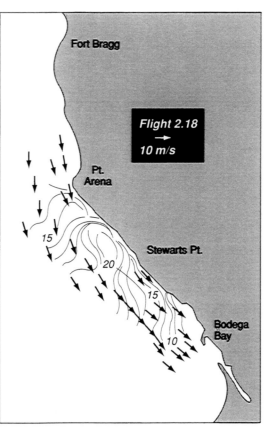

FIGURE 4.5 Isotachs ob-
served July 14, 1982, off central California from aircraft.
The increase in winds from
~15 m/sec to 20 m/sec is attributed to a decrease in inversion height from 300 to 50
m, which counters the deceleration associated with the
change in coastline orientation. The decrease from ~20
m/sec to 10 m/sec is associated with a hydraulic jump
with the inversion base quickly
increasing to 200 m (after
Winant et al., 1988).

can accelerate down the mountain slope into a fjord and continue over the
ocean. Changes in the orientation of the coastline can produce eddies, such
as those near Cape Mendocino (Dorman, 1985; Reason and Steyn, 1990) or
local changes in inversion height (Figure 4.5) with a local wind maximum
on the scale of 30 km, such as those south of Point Arena (Samuelsen,
1992; Winant et al., 1988).

The katabatic wind is a drainage wind current caused by the gravita-
tional flow of cold air off high ground, whose direction is controlled almost
entirely by orographical features. The strongest and most frequent katabatic winds occur in high latitudes, most notably Antarctica (see Bromwich,
1989; Ohata et al., 1985; Parish and Bromwich, 1989). Other high- and
midlatitude katabatic winds have been reported in Greenland by Gryning
and Lyck (1983); in Brugge, Belgium, by Dawe (1982); and in coastal
Alaska by Reynolds (1983) and Macklin et al. (1988). Katabatic winds in
low latitudes (Veracruz, Mexico) have been described by Fitzjarrald (1986).

Ackerman (1982) has summarized observations in Poona, India, in Mauna Loa, Hawaii, and in Cato, South Africa. Lopez and Howell (1967) have described katabatic winds in the equatorial Andes.

Coastal katabatic winds prevent fog formation, or when fog is present, they may clear it. Although katabatic winds typically dissipate within a few kilometers downstream of the forcing, Bromwich and Kurtz (1984) found that katabatic winds inland of Terra Nova Bay, Antarctica, are present some 25 km beyond the end of the main slope; this is also seen for southeastern Alaska in Figure 4.6. The physical processes important in the development of katabatic winds are radiative cooling, development of a sloped pressure gradient force, adiabatic warming of downward moving air, friction, and entrainment (Gutman, 1983; Manins and Sawford, 1979; Nappo and Rao, 1987; Parish, 1984; Parish and Waight, 1987).

The forcing of katabatic winds can be traced to the strong cooling of air adjacent to the surface. The net long-wave radiation is initially the dominant forcing term, acting to cool the surface of the slope. According to

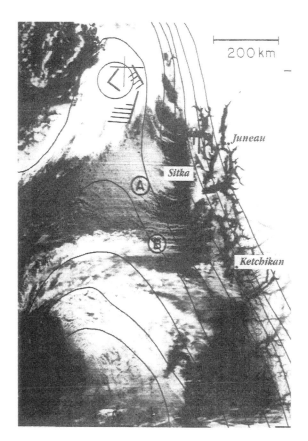

FIGURE 4.6 Satellite photograph of the Gulf of Alaska in winter. As a low-pressure system approaches the coast, strong, cold ageostrophic winds exit the fjords and continue across the continental shelf, as indicated by the low-level cloud streets. These winds merge with the synoptic flow either gradually, as in this case, or with a frontal feature. Such coastal wind systems influence coastal-ocean circulation through increased wind stress and wind stress curl (from Gray and Overland, 1986).

Lettau and Schwerdtfeger (1967), one serious limitation on the continued occurrence of katabatic winds in a region is the supply of cold air upslope. Persistent katabatic flow requires replenishment of cold air either by convergence of drainage currents upwind or by continued localized intense cooling. As katabatic winds develop, the mixing processes become enhanced, and the turbulent heat flux to the surface increases. This efficient removal of heat from the lowest levels of the atmosphere produces the sloped-inversion pressure gradient force and leads to further enhancement of katabatic winds.

The depth of the katabatic layer is on the order of 100 m, although depths as great as 1000 m may be found on the Antarctic coast. The coastal katabatic winds effectively mix a deeper atmospheric column, thereby spreading the diabatic cooling throughout the boundary layer. As a result of this mixing, air in the free atmosphere must be entrained into the katabatic layer. This entrainment process acts as a retarding mechanism on the katabatic flow by warming the katabatic layer and by acting as a drag on it.

Although a reasonable understanding of katabatic winds has been obtained, gaps in our understanding still remain. Two areas of uncertainty are the mechanisms of radiative cooling and turbulence in sloping flow, including entrainment. According to Manins and Sawford (1979), direct computation shows that the divergence of radiation responsible for cooling the air is much smaller than the surface radiation flux used in models of katabatic winds. This raises the issues of how deep the cooling layer is and whether water hazes or fogs play a role in radiative cooling. During rapid nocturnal cooling, condensation may occur, affecting radiative transfer. Particularly in polar regions, formation of low-level ice crystals in the cooling air appears to be the rule rather than the exception (see, e.g., Bromwich, 1988; Curry, 1983; Curry et al., 1990; Overland and Guest, 1991), and their presence has the potential to substantially perturb the radiation balance. Also, the radiative effect of low-level clouds within the katabatic layer has not been investigated; cloud-top radiative cooling from low-level clouds may enhance inversion-slope pressure gradients in the katabatic layer even though actual surface cooling may be reduced, while higher clouds will simply retard surface cooling and not contribute to the pressure gradient. Improved observations are required to elucidate the fine-scale structure of the katabatic flow.

To date, the modeling of katabatic winds has focused on understanding the physical mechanisms rather than actually predicting their occurrence and strength. Prediction of katabatic winds will require that consideration be given to the entire katabatic setting, including the dynamic and thermodynamic processes occurring upslope.

SUMMARY AND CONCLUSIONS

The major feature of the interaction of synoptic-scale flow with coastal topography is that mesoscale features are formed that have the scale of the half-width of broad coastal mountains or, more typically, for steep mountains have the scale of the Rossby radius based on mountain height. These scales of 50 to 150 km are smaller than observing networks, especially over water. Their dynamics are qualitatively known, and local orographic forcing is known by specifying the fine-scale topography. In principle, it should be possible to forecast coastal winds by using a regional numerical model driven by the larger synoptic-scale flow and the coastal orography. Inadequately studied areas limit current understanding of orographically dominated coastal meteorology. To address these limitations the panel recommends the following:

• Case studies that include the low-level wind regime should be undertaken to explain the modification of the structure and path of storm systems by coastal orography.

• A climatology should be developed of the synoptic-scale regimes that initiate or terminate coastal-trapped phenomena and the type of feature, wave-like or front-like, that evolves from each regime.

• Research to improve the resolution of numerical weather forecasting models must be conducted to focus on scales distinguishing important coastal processes.

• Studies should be conducted to determine the predictability of small-scale hydrodynamic features, such as hydraulic jumps, eddies, wind maxima, and katabatic flows.

• Long-term measurement programs should be supported to determine the low-frequency climatology of winds in the coastal zone. Ideally, observing system operations should be coordinated with mesoscale atmospheric models running on a continuous basis.

5

Interactions with Larger-Scale Weather Systems

As traveling disturbances pass over a coastline or ice edge, they experience a change in surface friction, heat fluxes, and possibly orography. The sudden change in surface conditions can modify the disturbance, but it also can give rise to entirely new phenomena that are peculiar to the coastal region. Similar effects can be produced indirectly, through interactions with the local coastal circulations discussed in preceding sections. Furthermore, variations in large-scale conditions can strongly influence coastal circulations, producing phenomena that are not found in steady-state large-scale conditions. The complexities introduced by such mesoscale and synoptic-scale interactions hinder conceptual understanding, and accurate forecasting of changing coastal conditions requires the simultaneous simulation of a variety of disparate processes and their interactions. Manifestations of these interactions are numerous; only examples are provided here.

LAND-FALLING HURRICANES

One example of a weather system that is directly modified by the change in surface conditions across a coastline is the typhoon or hurricane. Hurricanes require a warm ocean for maintenance and intensification. Although hurricanes soon weaken over land, the weakening is accompanied by potentially devastating short-term phenomena. For example, orographic barriers influence the dynamics of land-falling hurricanes through the blocking and resultant disruption of the circulation of the storm. Torrential rainfall in coastal environments and large latent heat releases result from this interaction. Even storm motion is influenced, as shown by Bender et al. (1987) for the island of Taiwan; there the storm moved north of the westward translation it would have had in the absence of the terrain.

During and after landfall, the near-surface winds associated with the storm decelerate as a result of increased surface friction, which causes an expansion in radius of the eye wall cloud as well as creation of a stably stratified atmosphere in the lower atmosphere due to lower tropospheric cooling. Above the lowest levels, however, the winds usually remain strong and can even increase temporarily as coupling with the surface is diminished. The resultant large vertical wind shear has been used to explain the frequent occurrences of tornadoes in these storms (Novlan and Gray, 1974, as described in Pielke, 1990, Figure 4.6). Some land-falling hurricanes, however, do not produce tornadoes. We need to understand better the complex interaction of the coastal environment and the hurricane that often, but not always, results in tornadoes (e.g., with Hurricanes Carla in 1961 and Celia in 1970). A first modeling study of deep cumulus convection in hurricane environments is reported by McCaul (1991). Farther inland the acceleration of winds aloft due to decoupling from the surface may explain localized areas of tornado and strong wind damage such as those that occurred in Charlotte, North Carolina (Hurricane Hugo, 1989), and Washington, D.C. (Hurricane David, 1979). Extratropical cyclones are also modified significantly as they strike land, as found, for example, on the Washington coast in the Cyclonic Extratropical Storms (CYCLES) project (Hobbs et al., 1980).

POLAR AND ARCTIC LOWS

The presence of a coastline or ice edge at high latitudes appears to strongly favor the formation of intense storms. Such storms, called arctic-front-type polar lows by Businger and Reed (1989), form over water just beyond the ice edge, where the large temperature contrast between cooler air and warmer water leads to strong low-level baroclinicity and weak stratification. As documented by Reed and Duncan (1987), trains of polar lows can form along the strong temperature gradient (arctic front), which suggests an instability mechanism. On the other hand, the most intense polar lows form rapidly beneath upper-level troughs that move from the ice shelf to the open ocean. According to Businger (1991) and Emanuel and Rotunno (1989), some polar lows attain the intensity and structure commonly associated with hurricanes.

The range of possible mechanisms for polar low formation includes barotropic and baroclinic instability, conditional instability of the second kind, and air-sea interaction instability (Twitchell et al., 1989). Recent evidence suggests that even katabatic winds (see Chapter 4) can play a critical role in the formation of some Antarctic lows (Bromwich, 1991). Progress has been hindered by the lack of observations on sufficiently small space and time scales over the lifetime of polar lows, although interesting case studies have recently emerged from the Arctic Cyclone Experiment (Shapiro et al., 1987) and the Ocean Storms Experiment (Bond and Shapiro, 1991). Although observations seem to indicate that more than one mechanism operates in individual polar low events, little is known at present about how the proposed mechanisms interact with each other.

HYBRID FRONTAL CIRCULATIONS AND WINTER STORMS

Along midlatitude coastlines, a wide range of processes are often present and influence each other. Important hybrid mesoscale circulations can develop that are a combination of thermally forced (Chapter 3) and topographically forced (Chapter 4) processes. One example is locally known in the northeastern United States as the backdoor cold front (Bosart et al., 1973), in southeast Australia as the southerly buster (Colquhoun et al., 1985), and in New Zealand as the southerly change (Smith et al., 1991). When the sea (or other large body of water) is much colder than the land, a cold front moving parallel to the coast undergoes local intensification and advances rapidly along the coast while being inhibited farther inland. While the cold front appears to be orographically trapped, numerical simulations have shown that differential surface heat fluxes play a primary role in modification of the cold front (Howells and Kuo, 1988).

Another type of hybrid circulation is common along the east coast of the United States during winter storms. As the wind becomes easterly ahead of a winter cyclone, coastal frontogenesis occurs, and cold air damming occurs along the mountains. After a few hours, the trapped cold air is bounded to the east by the coastal front, whose inversion extends along the top of the cold dome toward the mountains. The combined circulation enhances the coastal front by maintaining a source of cold air and keeping the front nearly stationary. As the coastal front intensifies, the temperature difference across the inversion increases, strengthening the damming and making it more difficult for the cold air to advect over the mountains. The combination of shallow cold air against the mountains, warm moist air ascending over the cold dome, and an approaching winter storm often leads to low visibility and hazardous ice storm conditions for sites east of the Appalachians (Forbes et al., 1987). Details of the interaction between the coastal front and cold air damming, such as whether one tends to trigger the other, are still unclear.

Coastal regions of the eastern United States are a favored location for cyclogenesis, and the resulting coastal storms tend to be rather different from their counterparts over the open Pacific and Atlantic or in the Midwest. East Coast cyclones have been shown to be generated or strongly modified by the Appalachian Mountains, the land-sea contrast, and the sea surface temperature contrast concentrated along the north wall of the Gulf Stream, as well as such coastally confined circulations as cold air damming and coastal fronts. (For examples, see Hobbs, 1987, and Holt and Raman, 1990.) The close geographical proximity of all these influences within the coastal region makes the individual interactions difficult to separate. In complex winter storms, such as the Presidents' Day storm of 1979, cyclogenesis has been found to depend on all the above factors acting in concert (Lapenta and Seaman, 1990; Uccellini et al., 1987).

LOCALIZED LATENT HEAT RELEASE

One modifying coastal influence not yet discussed is localized latent heat release. In the prestorm environment, cold dry continental air has typically been advected from the northwest. This air remains cold and dry over land, while over water it receives both heat and moisture from the sea surface in a process known as preconditioning. When a developing storm approaches, the extra energy ingested by the marine boundary layer encourages rapid intensification of the storm over water or along the coast. The heat fluxes that occur as the storm intensifies are generally less important (Kuo et al., 1991).

When a strong cyclonic circulation is not already present, any mechanism that tends to focus latent heat release has the potential to produce a small-scale cyclone. The dynamics of coastal cyclones and low-level jets that form along the Baiu (Mei-yu) front of eastern China and Japan appear to be strongly dominated by the direct effects of latent heat release focused by orography or a frontal wave (e.g., Chen and Yu, 1988; Nagata and Ogura, 1991). Along the east coast of the United States, it has been suggested that coastal fronts can supply the necessary focusing mechanism and cooperatively interact with the resulting cyclone. For example, Keshishian and Bosart (1987) documented a case in which a small-scale coastal cyclone propagated northeastward along an extensive coastal front. Because the circulations associated with the coastal low acted to enhance the temperature gradient and strengthen the coastal front, it was called a "zipper low." The increased low-level baroclinicity and low-level moisture may have played an important role in the major cyclogenesis event that followed.

The triggering mechanism for coastal latent heat release may be large-scale upward motion or forced lifting by a coastal front and cold air damming. In either case, a large-scale trough is generally approaching from the west to provide the necessary flow configuration. The combination of large-scale ageostrophic circulations and forced lifting is thought to have caused the initial formation of the Presidents' Day low, but small-scale coastal cyclogenesis has sometimes been triggered by large-scale ascent alone in the absence of a coastal front or cold air damming.

SUMMARY AND CONCLUSIONS

There are large gaps in our present understanding of coastal interaction processes. Theoretical studies tend to focus in isolation on specific processes that are most amenable to analytical treatment and interpretation. By contrast, observational and numerical studies tend to focus on the most extreme cases, and it has been found that the extreme cases tend to involve a wide range of mesoscale and synoptic-scale processes interacting cooperatively. The range of interactions also presents a modeling challenge, since successful forecasts or simulations must adequately handle both the individual processes and their interactions. Without a better understanding of the nature of the interactions, verification of model dynamics and improvement of numerical forecasts become difficult.

We recommend strongly focused numerical, observational, and theoretical investigations into specific mesoscale-synoptic interactions:

• Research should be conducted to determine the dynamics of the local intensification of cyclone winds by coastal topography and the resulting modification of storm intensity and motion.

• Further research should be implemented to identify the causes of strong local winds, tornadoes, and extreme precipitation within land-falling hurricanes, polar lows, and extratropical cyclones.

• Studies should be conducted to understand the role of the coastal baroclinic zone and katabatic winds in the formation and dynamics of polar lows.

• Studies should be undertaken to determine the dynamical influence of coastal heating discontinuities in the along-shore propagation and local intensification of cold fronts.

• Studies should be carried out to understand the role of topography in the formation and motion of coastal fronts.

• Further research to quantify the nature of the influence of coastal fronts on midlatitude coastal cyclogenesis should be supported.

• Research should be conducted to clarify the importance of coastally induced moisture inhomogeneities for small-scale cyclogenesis and low-level jet formation.

6

The Influence of the Atmospheric Boundary Layer on the Coastal Ocean

Physical processes that affect the coastal environment are often unique because of the dominant effect of the interactions between the land and the sea. The thermal contrast between the land and the sea contributes to the formation of the land-sea breeze, coastal atmospheric fronts, and atmospherically induced coastal ocean currents and upwelling. The presence of a lateral boundary can substantially modify the wind field that drives currents, upwelling, and surface waves, creating highly variable fields. As a consequence, the processes that drive the air-sea exchange of heat, mass, momentum, and trace gases over the continental shelf are highly variable; they are dominated by spatial scales on the order of tens to hundreds of kilometers. Stratiform clouds, which affect the radiation balance significantly, can form over cooler upwelled water (e.g., Rogers and Olsen, 1990), while the convergence of marine air over the coastline can result in strong convection with heavy precipitation and runoff. These mechanisms affect pollutant dispersion, coastal erosion and beach development, the coastal ecological system, and numerous other processes.

The importance of air-sea interaction processes to the flow and thermodynamic structure of the ocean is well established (Charnock, 1979). The wind-driven circulation on the continental shelf is controlled by local wind forcing on time scales on the order of hours and by remote wind forcing on time scales on the order of days. A substantial fraction of the variability of coastal currents, sea level, and temperature results from large-scale ocean waves that are coherent with remote wind forcing and can propagate many hundreds of kilometers (Davis and Bogden, 1989; Denbo and Allen, 1987).

The lateral boundary is an important element in the coastal circulation on all scales because it can constrain the flow in both the atmosphere and the ocean, and there is often close correspondence between changes in the currents and the local wind forcing.

The coastal ocean is characterized by large variations in sea surface temperature and roughness and a nonequilibrium sea state. Upwelled water is often much colder than the ambient surface water, so sharp ocean fronts form between the colder water nearshore and the warmer water offshore. These regions represent areas of particularly intense air-sea interactions because of large inhomogeneities and nonequilibrium conditions. The structure of the atmospheric boundary layer becomes increasingly complex in the vicinity of ocean fronts. Changes in the heat flux produce large variations in the stability of the boundary layer, cloud cover, and radiative fields on scales up to 100 km. Recent atmospheric measurements in the vicinity of a weak ocean front in the open ocean have revealed complex wind stress, cloud cover, and boundary layer depth patterns (Friehe et al., 1991). More dramatic effects are likely in the coastal ocean where fronts are persistent and strong (Charnock and Businger, 1991). The high biological productivity of the coastal environment also emphasizes the importance of understanding trace gas exchange in the presence of these large horizontal gradients. In particular, the coastal ocean may be an important sink for carbon because the surface partial pressure of CO_2 is controlled by photosynthesis, which depletes carbon from the upper layers (Baes and Killough, 1985; Broecker, 1982; Sarmiento et al., 1988; and a review by Broecker et al., 1985). Present models of the CO_2 cycle, however, tend to be limited to larger spatial scales and do not distinguish the effect of the coastal ocean (Moore and Björkström, 1986; and a review by Crane, 1988). In addition, it has been postulated that dimethyl sulfide, produced by phytoplankton, may be an important source of cloud condensation nuclei (Bates et al., 1987; Charlson et al., 1987; Hegg et al., 1991).

Recent measurements have also shown that the response time of currents on the continental shelf is sufficiently short that changes in ocean circulation patterns can be driven by strong wind events, such as those caused by the passage of a strong atmospheric front (Lee et al., 1989) or the interaction between mesoscale or regional pressure gradients and topography (e.g., Lackmann and Overland, 1989), and by intense cooling of the upper ocean associated with cold air outbreaks (e.g., Bane and Osgood, 1989). The problem is complicated further by mesoscale circulation patterns, like the sea breeze, which contribute to the small-scale variability of wind stress (Clancy et al., 1979; Elliot and O'Brien, 1977; Mizzi and Pielke, 1984), and by open ocean processes such as the Gulf Stream that can force circulation on the outer continental shelf (Lee et al., 1989). Interactions between the atmosphere and the coastal ocean are characterized by their intensity and horizontal variability on very small scales. The lower atmosphere, in general, responds quickly to changes in sea surface temperature and roughness. Where upwelling is particularly intense, the cold near-shore water cools the atmosphere, forming a shallow stable marine layer, capped by a large temperature inversion. Often this layer is below the height of the coastal terrain, so the marine air behaves in a manner that is very sensitive to the orientation of the coastline (Winant et al., 1988). A complete understanding of the effect of the marine atmosphere on the circulation and thermal structure of the coastal ocean requires a better understanding of how the air and sea interact on scales less than 10 km and less than 1 hour. A key to this understanding is realization of the possibility of strong coupling between the ocean and the atmosphere.

COASTAL PROCESSES

Local and Remote Wind Forcing

An important consideration in the coastal environment is the relative contribution of local and remote wind forcing to the ocean circulation. Remote forcing consists of the generation of coastally trapped long waves that have wavelengths on the order of 1000 km. The local wind velocity consists of two components, one part driven by the pressure gradients and another part driven by the stress divergence, which is confined to a shallow layer called the Ekman layer. Along the west coast of North America, the response of sea level to fluctuations in along-shore wind stress at large scales accounts for a substantial fraction of the total sea-level variance. Halliwell and Allen (1984, 1987) have shown that the most effective wind stress forcing was confined to two regions, along the northern California and Oregon coasts, and northern Baja, California. The strongest wind-stress-forced fluctuations in sea level propagated northward, with the maximum correlation between sea-level fluctuations and the wind stress having lagged about 500 km equatorward 1 to 2 days earlier. Thus, large-scale processes play a major role in determining the wind-driven shelf circulation. Variance of the wind stress is largest near capes (Enriquez and Friehe, 1991; Halliwell and Allen, 1984), where strong local forcing may trigger wind-driven large-scale coastal-trapped waves.

One important coastal phenomenon is the wind-driven cross-shelf circulation that drives upwelling and downwelling (Figure 6.1). An equatorward along-shore wind on a west coast produces an offshore surface flow driven by turbulent stresses (Ekman layer). This offshore flow is compensated for by an onshore flow deeper in the water column. In addition to the horizontal cross-shelf currents generated by the wind stress, there is a compensating vertical velocity pattern that results in irreversible incorporation

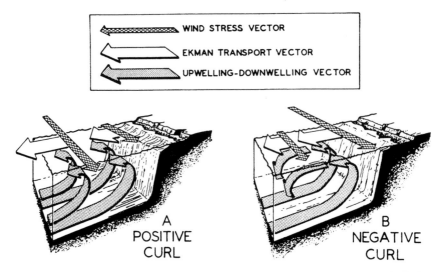

FIGURE 6.1   Relationship of along-shore winds and coastal upwelling and of wind stress curl and divergence/convergence of surface Ekman transport offshore of the primary upwelling zone (from Nelson, 1977).

of cold nutrient-rich deep water into the surface layer. The dynamics of the return flow that feed the Ekman layer are not well understood. An important question is whether this return flow could be geostrophic. For the CODE area, Davis and Bogden (1989) have shown that the pressure gradient is inconsistent with a geostrophic return flow to balance the Ekman transport, so that some other ageostrophic component is indicated. This suggests that the effect of the topography and local thermally driven atmospheric circulation on the wind flow may be particularly important in determining the wind-driven circulation. This problem, however, has hitherto not been addressed.

Ocean Fronts

In a wind-driven upwelling environment, a sharp discontinuity in density typically develops between the less dense surface water offshore and the denser upwelled water near shore. Although an along-shore wind stress is essential to the formation of a coastal upwelling front, the effect of the wind stress on the stability of a front is uncertain. The atmosphere may respond to this ocean thermal structure on very small scales (<10 km), thereby affecting the wind speed, stress, and heat exchange between the atmosphere and ocean. In turn, there may be further feedback from the atmosphere to the ocean, also indicating that the coastal ocean and atmosphere may be strongly coupled. Friehe et al. (1991) have shown that the wind and wind stress field can vary dramatically in response to the atmospheric stability differences that exist across an ocean front. The result is a highly variable time-dependent stress pattern that alters the basic flow field in the oceanic Ekman layer. Interactions with this flow are likely to affect the stability of a coastal upwelling front. Thus, time-dependent changes in the wind field and the Ekman layer should be considered in studies of upwelling and coastal frontal dynamics.

Atmospheric forcing also affects the structure of the temperature and salinity fronts that develop between cool shelf water and the western boundary currents that flow adjacent to the eastern continental shelves of most continents at midlatitudes. This large temperature or salinity discontinuity is maintained by momentum and buoyancy fluxes across the air-sea interface and by changes in the path of the boundary current. Rapid cooling of the sea surface may intensify the frontal structure because the shallow shelf water cools at a faster rate than the water offshore, which, in any case, is resupplied with heat by the poleward advection of warmer water. There is substantial feedback from this warm current to the atmosphere, where storms can develop, resulting in further interaction with the ocean. This is discussed later in more detail. Air-sea interactions also modify frontal boundaries that develop in buoyancy-driven and tidally dominated coastal environments where the heat exchange between the air and the sea affects the strength of the front. Of particular importance is the effect of episodic events, such as cold air outbreaks and hurricanes.

Storms

Severe storms affect many coastal regions during the winter and, to a lesser extent, in summer. Regional differences exist, with ocean storms affecting the coast from California to Alaska and extratropical cyclones moving from the interior of the continental United States to the east coast. Tropical cyclones, including hurricanes, may affect almost any part of the coast during summer and autumn. Particularly important effects are caused by cold air outbreaks along the east coast and by rapidly intensifying storms over the Gulf Stream.

Cross-shelf transport off the eastern coast of the United States is substantially affected by a combination of wind and Gulf Stream forcing (Lee and Atkinson, 1983; Lee et al., 1989). The atmospheric effect is most significant during cold air outbreak episodes in winter. This atmospheric forcing is strongly influenced by the large gradients in surface temperature that lead to low-level frontogenesis (Bane and Osgood, 1989). This is accompanied by substantial changes in the surface stress that drives the cross-shelf Ekman transports in the surface layer. The response time of the shelf currents is sufficiently short that changes in the shelf circulation can be driven by rapidly moving atmospheric events (Lee et al., 1989). The response time of the inner- and middle-shelf regions in these conditions is between 3 and 12 hours.

Topography

Interaction between the regional or mesoscale pressure gradients and topography leads to complex wind and turbulence fields over the coastal ocean (Macklin et al., 1988; Overland, 1984; Chapter 4, this report). Strong ageostrophic winds occur in topographically restricted channels when synoptic-scale disturbances produce an along-channel pressure gradient (Lackman and Overland, 1989). These gap winds produce a highly variable surface stress field through local enhancement of the flow downstream of the topographic gap. Large variations in the wind field and corresponding changes in the surface pressure over short spatial scales have been observed along the coast of California (Winant et al., 1988; see Figure 4.5). This pattern, which occurs with generally northerly winds, is characterized by a strong low-level temperature inversion at the top of the marine atmospheric boundary layer, and the spatial structure of the surface wind is correlated with the coastal topography.

There have been extensive studies of topographic effects on the development of internal waves in a stable stratified fluid; however, the effect of topographically forced gravity waves on the coastal marine atmospheric boundary layer and sea surface layer has received relatively little attention (Dorman, 1987; Gossard and Munk, 1954; Wald and Georgopolous, 1984). In a stable coastal environment, mesoscale variations in the flow pattern may be dominated by topographically forced waves that propagate offshore, in the lee of elevated coastal terrain, or that propagate parallel to the coast as trapped waves. Gossard and Munk (1954) observed a slight effect of a propagating atmospheric gravity wave field on sea level using measurements of surface pressure and sea level from the pier at Scripps Institution of Oceanography. Interaction of atmospheric waves with the sea surface was observed by Wald and Georgopolous (1984) using Advanced Very High Resolution Radiometer (AVHRR) satellite data. Alternate light and dark features seen downwind of islands in the Aegean Sea were accounted for by changes in the sea surface temperature and roughness. Reversals in the wind direction and variations in the depth of the marine layer along the west coast of the United States have been related to gravity current surges along the coast (Dorman, 1987). The effect of the sea surface temperature and stress patterns is uncertain, but the observations of Enriquez and Friehe (1991) indicate that sea surface temperatures along the northern California coast respond quickly to changes in reversals in wind direction due to changes in the upwelling pattern. Herman et al. (1990) have shown that a drop in sea surface temperature of 1° to 3°C from 15 km offshore out to approximately 150 km is accompanied by a rise of similar magnitude within approximately 15 km of the shore in response to highly nonlinear bores or gravity currents.

AIR-SEA EXCHANGE PROCESSES

The exchange of heat, mass, moisture, momentum, trace gases, and particulates between the sea and the air is fundamental to an understanding of the ocean-atmosphere system and global climate. Although such processes are ubiquitous over the ocean, the coastal environment poses certain difficulties. The land-sea boundary, ocean temperature fronts, surface wave field, variations in water depth, and biological activity play an important role in determining the magnitude and variability of air-sea fluxes over the coastal ocean.

Wind Stress, Heat Fluxes, and Trace Gas and Particulate Exchange

Momentum flux is particularly important in determining the effect of the atmosphere on coastal circulation. Direct measurement is generally difficult, so drag coefficients that relate the turbulent stress to the mean wind are often used (Fairall and Larsen, 1986; Smith, 1988). Accurate estimates of the exchange coefficients of momentum between the atmosphere and the ocean depend strongly on atmospheric boundary layer stability, wind shear, and the ocean surface wave field (Donelan et al., 1985; Smith et al., 1990a). This is further complicated by the limited fetch and water depth in the coastal zone, and the heterogeneity of the coastal atmosphere that limits application of the steady-state assumptions inherent in the formulation of bulk parameterizations of the surface exchange processes (Geernaert, 1988, 1990; Chapter 2, this report).

The estimate of transfer velocities for trace gas exchange at the air-sea interface remains difficult in the ocean and is further complicated in the coastal environment by the variability of the wind and temperature fields and the structure of the upper ocean. Estimates of transfer velocities can be obtained using various methods (for a review, see Liss, 1988).

Assuming that production of natural $^{14}CO_2$ in the atmosphere is time invariant and that a steady state exists in the ocean-atmosphere system, the rate at which $^{14}CO_2$ enters the oceans across the air-sea interface must balance the rate at which ^{14}C decays in the oceans, and a global estimate of the transfer velocity can be obtained (Broecker and Peng, 1974; Liss, 1988). Bomb-produced ^{14}C can also be used to estimate the air-sea transfer velocity. Another method uses the change in oxygen concentration obtained from a time series of dissolved oxygen. The change in concentration is attributed to biological production and consumption due to photosynthesis-respiration and decomposition and exchange across the air-sea interface. The biological component is also related to nutrient changes, so the oxygen concentration can be corrected for biological activity. The residual signal is then attributed to the transfer velocity for oxygen. Radon deficiency has also been used. This method relies on measuring the radioactive disequilibrium between the nuclides ^{222}Rn and ^{226}Ra in near-surface water. In the deep oceans the two isotopes are found to be generally in radioactive equilibrium. Near the surface, however, the gas ^{222}Rn is lost to the atmosphere because its activity in air is much less than in seawater. By assuming steady state, for any measured vertical profile the rate loss of ^{222}Rn to the atmosphere must equal the depth-integrated deficiency in ^{222}Rn activity, and a transfer velocity can be estimated. None of these methods attempt to estimate the flux of a tracer across the air-sea interface directly; all of them depend on steady-state assumptions that are difficult to relate to environmental parameters such as wind speed, sea state, and trace gas concentrations in the surface water, all of which vary over short time scales. Ideally, direct flux estimates are needed that are based on atmospheric measurements of the covariance of the vertical velocity and trace gas concentration in the atmosphere. This method has been used to estimate the air-sea flux of CO_2 (Smith and Jones, 1985; Wesley et al., 1982), but the results are controversial because they give a transfer velocity for CO_2 that exceeds the oceanographic estimates by a factor of 100. A significant problem in the eddy correlation approach is the high signal-to-noise ratio where direct estimates of the variance of CO_2 are required. This has been overcome recently with a more sensitive CO_2 sensor (S.D. Smith et al., 1991). These observations indicate upward and downward fluxes of CO_2 in response to daily variations in CO_2 in the coastal surface seawater where there are changes in the salinity of the surface water due to freshwater runoff. Transfer velocity estimates are an order of magnitude larger than the open ocean measurements made using radon but are similar to the wind tunnel and lake studies of Liss and Merlivat (1986). Recent advances in chemical flux estimates using conditional sampling techniques that combine estimates of the variance of the vertical velocity and the mean concentration of the trace gas may reduce much of the uncertainty in atmospheric measurements of tracer fluxes across the air-sea interface (Businger and Delany, 1990; Businger and Oncley, 1990).

Coupled Interactions with the Planetary Boundary Layer

Coupled air-sea interactions occur when, for example, the atmosphere is forced by the ocean, resulting in a feedback to the ocean that modifies further response of the atmosphere. Coupling occurs throughout the ocean-atmosphere system; however, it is more pronounced in the coastal environment where the temporal and spatial scales in the ocean and the atmosphere are similar. While surface exchange processes directly control the interaction of the ocean and the atmosphere, they depend greatly on the mesoscale structure of the entire boundary layer. The sea breeze, the interaction of a stable boundary layer with topography, boundary layer rolls, and clouds may all contribute to the variability of the surface fluxes. The importance of the sea breeze for enhancement of the surface wind field and subsidence offshore has been demonstrated along the Oregon coast (Clancy et al., 1979; Elliot and O'Brien, 1977; Mizzi and Pielke, 1984). Feedback to the mesoscale circulation is likely if surface temperature gradients are modified by the sea breeze circulation. Modification of the surface stress field in the vicinity of a cape along the California coast has recently been observed in the Surface Mixed Layer Experiment (SMILE) (Enriquez and Friehe, 1991). The structure of the boundary layer may be further modified by radiative processes associated with cloud development, resulting in variations of the surface fluxes due to changes in the vertical structure of the marine atmosphere (see Chapter 2).

The horizontal variability of surface temperature fields and land-sea contrasts leads to modification of the air and development of internal boundary layers (IBLs). As air flows over an abrupt change in surface properties, an internal boundary layer develops within an existing boundary layer. For the case of cold air flowing over a warm surface, an unstable IBL develops, rapidly replacing the existing boundary layer; for the case of warm air flowing over a cold surface, a stable IBL develops that may persist as a shallow layer until either the air is cooled by radiation or the surface temperature increases to reverse the stability of the air. This stable boundary layer accounts for many of the more important effects that occur in the interaction of the marine boundary layer with topography (Dorman, 1987; Winant et al., 1988).

SUMMARY AND CONCLUSIONS

It is clear that the coastal ocean and atmosphere are inherently interdependent or coupled because of the dependence of the atmosphere on the heat and moisture source at the sea surface and the dependence of the ocean circulation on the wind. Traditionally, however, studies have rarely combined investigations of both environments to determine the extent of the feedback between the ocean and the atmosphere or the extent to which these fluids are coupled. This applies also to the exchange of trace gases between the atmosphere and the ocean that depends on the structure of the upper ocean and atmospheric boundary layer. Direct interaction between the atmosphere and biological or geological processes in the ocean is more tenuous; however, the physical interaction between the atmosphere and the ocean modifies the cross-shelf circulation and therefore plays an important, albeit indirect, role.

Studies of the coupled ocean-atmosphere system in the coastal environment depend largely on understanding the scales of interaction between the two fluids and the processes that provide the strongest feedbacks. A focus on small-scale interactions would provide the opportunity to elucidate important physical processes that control the coastal ocean and atmospheric circulation and the larger-scale fields. A number of basic air-sea interaction problems are of particular interest. Parameterization of the surface fluxes in shallow water remains a problem. The surface wave field on the shelf is substantially different from that over the open ocean, and the effect of this on air-sea exchanges is not well known (Geernaert, 1990). The relative effect of topographically forced flows and thermally driven circulation on the coastal ocean circulation has not been explored. There is compelling evidence for both topographically forced and sea-breeze-driven wind stress variations along the west coast of the United States, although little is known about the effect of the thermally driven circulation on the structure of the marine layer that controls the hydraulically driven flow near the ocean surface.

The high wind speed zone in the lee of capes and points is associated with a large wind stress curl pattern (see Figure 6.1) and lower sea surface temperature in the vicinity of the cape. It is unclear whether this is due to increased surface cooling from the high winds, increased upper ocean mixing, or upwelling at small spatial scales. Similarly, these capes appear to be sources of the largest wind stress variability that propagates along the coast and may be responsible for the remote forcing of sea level and the coastal circulation. However, the relative importance of the local and remote wind forcing of the coastal ocean remains uncertain.

Another uncertainty is the effect of the increased surface fluxes associated with cold air outflow from the continent on coastal frontogenesis over the coastal ocean. At present, it is unclear whether the development of coastal frontogenesis is related to variations in the shelf circulation, larger-scale atmospheric forcing, or a combination of both. The extent of the feedback from coastal frontogenesis to the ocean via variations in the wind stress also is uncertain. The growth and decay of the atmospheric or oceanic boundary layer also are not well understood when internal waves, shear flow instabilities, and temperature fields are spatially highly variable.

Advances in our understanding of coastal phenomena will benefit from research to understand interactions between the ocean and the atmosphere over a large range of spatial and temporal scales. While it is apparent to oceanographers that the wind-driven ocean circulation is of fundamental importance on the continental shelf, it may be less clear to meteorologists that the highly heterogeneous state of the coastal ocean, in association with the complex topography, produces a highly variable atmospheric structure on very small spatial and temporal scales across the continental shelf. Efforts to understand the extent of the coupling between the atmosphere and the ocean in this highly variable environment will be very rewarding, providing new insight into the basic physics of the atmosphere and the ocean and improving our ability to predict the mesoscale ocean and atmosphere circulations.

In conclusion, the panel recommends the following:

• Studies should be conducted to determine the relative effects of topographically forced flows and thermally driven circulation on coastal ocean circulation.

• Further research should be conducted to understand the coupled ocean-atmosphere processes that control the interactions between the wind field, atmospheric boundary layer structure, and upper ocean.

• Support should be given to encourage development of coupled ocean-atmosphere models that use integrated field measurements in the coastal ocean and atmosphere.

7

Air Quality

A substantial fraction of the nation's and the world's populations, fossil and nuclear energy production, heavy industry, vehicular traffic, and energy exploration and production lie within 100 km of a coastline. The large emissions of primary, and the production of secondary, pollutants combined with complex and often adverse meteorological conditions can result in increased air pollution levels. An adequate understanding of coastal zone air pollution dispersion is required to license and control point sources of primary pollutants, design regional emission control strategies for secondary pollutants contributing to photochemical oxidants, develop and implement emergency responses to accidental releases of radiological and hazardous materials, assess dry and wet deposition of trace metals and other contaminants into sensitive coastal ecosystems, and, potentially, predict dispersion of militarily significant chemical and biological agents.

DISPERSION IN THE COASTAL ZONE

Substantial progress has been made in understanding the mechanisms involved with dispersion over distances ranging from tens of meters to circumglobal. However, a gap appears in our understanding of processes involved in mesoscale dispersion (10 to 100 km), particularly when mountainous and coastal regions are involved.

Coastal zone dispersion issues have been summarized by Lyons (1975), Lyons et al. (1983), Pielke (1984), and Zanetti (1990). Dispersion is the combined impact of diffusion and transport. Both are influenced by the intense gradients of surface fluxes, roughness, turbulence, insolation, mixing depths, and horizontal and vertical winds of the coastal zone. Modification of stable onshore flowing air over heated land results in sharp discontinuities in diffusive behavior, including continuous fumigation of elevated plumes (Lyons and Cole, 1976). A comparison of different coastal fumigation models is given by Stunder and Raman (1986). Plume trapping over land and ribbon plumes over water can result in locally high pollutant concentrations (Lyons et al., 1983). The diurnal cycling from land breeze to sea breeze presents special problems, especially for emergency response planners. Tracers, tetroons, sulfate aerosol data, and aerometric studies in the Los Angeles basin, the Chicago-Milwaukee area, along the U.S. east coast, and elsewhere suggest that materials transported offshore in a land breeze can advect back onshore in relatively high concentrations during the subsequent sea breeze and often at undetermined locations (Shearer and Kaleel, 1982).

Figure 7.1 summarizes many known coastal zone dispersion features and, rather like much of our understanding of these issues, is biased toward the two-dimensional and steady state. Yet plume transport in coastal zones involves inherently three-dimensional dynamic processes. Lyons and Cole (1976) hypothesized that mesoscale transport in sea breezes took the form of broad quasi-helical vortices extending parallel to the shore. Very fine-mesh mesoscale numerical dispersion modeling of complex wind fields at the Kennedy Space Center (Lyons et al., 1991a) demonstrated that even a dense two-dimensional surface layer wind network cannot consistently describe plume behavior. During sea breezes, plumes from continuous sources can evolve into contorted patterns.

As illustrated in Figure 7.2, pollutants moving inland and perhaps undergoing fumigation can thereafter be vertically translocated many hundreds of meters aloft in the strong updrafts associated with wind field discontinuities. Subsequently, these plumes can divide into several branches, with some material entering the gradient wind layer aloft and exiting the region. The remainder subsides into the sea breeze inflow layer, perhaps undergoing fumigation to the surface again many tens of kilometers from the source. These processes can become even more complicated in the presence of heated coastal mountain slopes.

Coastal katabatic winds are important locally since they result in the ventilation and consequent abatement of nocturnal pollutants in cities located on slopes (Ulrickson and Mass, 1990a, b). The size sorting of aerosols and pollen in regions of strong land and sea breeze systems' ascent and descent (Lyons, 1975; Lyons et al., 1991b) shows that shear-induced diffusion and dry deposition processes over water still require much additional study. The influence of coastal mesoscale systems on optical and other electromagnetic transmissions is not completely understood.

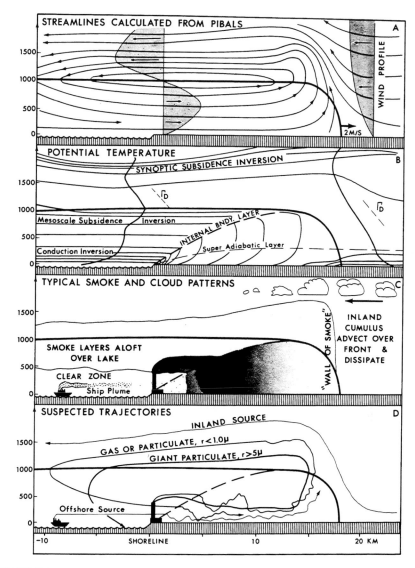

FIGURE 7.1 Structure of a typical lake or sea breeze circulation along a straight shoreline with flat terrain. Illustrated are recirculating flows, limited mixing depths in coastal zones, dynamic fumigation associated with the thermal internal boundary layer, and size sorting of aerosols. This two-dimensional conceptual model has been extant since the mid-1970s (Lyons, 1975). Current efforts require verifying and quantifying many aspects of this model and expanding it to include the three-dimensional structure of the land-sea breeze and its impacts on mesoscale transport and clouds in regions of irregular coastlines, islands, estuaries, and complex topography.

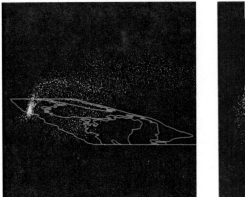 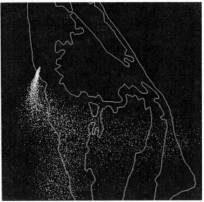

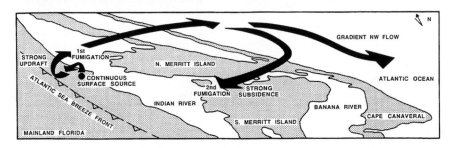

FIGURE 7.2 (top) Dispersion model simulation of plume transport in the complex coastal wind fields of the Kennedy Space Center. Plan view (left) of the predicted plume of particles released from a continuous surface source over the Indian River, and (right) the same plume seen from an elevated southwesterly perspective showing the three-dimensional nature of the transport processes. (bottom) Schematic showing that pollutants initially move inland in the Atlantic sea breeze inflow. They then rise rapidly in the strong updrafts in the sea breeze front over the Florida mainland. Some of the pollution subsides back into the sea breeze inflow due to strong subsidence over the Indian River and then fumigates again as it moves inland some 20 km south of the source. The other branch of the plume, ejected higher into the gradient flow, drifts almost due east as it exits the region (Lyons et al., 1991a). Graphics courtesy of Cecil S. Keen.

The mechanisms by which vertical and horizontal recirculation processes degrade coastal air quality are only partially understood. Pollutant reentrainment back into the inflow from the elevated return flow layer in offshore subsidence zones remains unquantified. Coastal circulations are major factors in exacerbating regional photochemical ozone pollution. Violations of the U.S. federal standard for ozone are four times more likely within 20 km of the Lake Michigan shoreline than inland (Lyons et al., 1991b). Worldwide ozone levels in cities as diverse as Oslo, Toronto, Athens, and Tokyo are strongly influenced by local sea breeze circulations and interactions with urban and complex terrain-induced circulations. Low-level jets in lake breeze cells over the Great Lakes have not been well documented but may play a crucial role in regional photochemical pollution transport. Developing improved conceptual and mathematical models of the myriad processes involved is an ongoing challenge.

DEVELOPING IMPROVED DISPERSION MODELS

The majority of diffusion simulation codes used today utilize the straight-line Gaussian plume (or its progeny), segmented plume, or puff models. This results largely from these models' relatively modest computational requirements, easily available input data, conceptual simplicity, and widespread acceptance by regulatory agencies. While they achieve modest accuracy in idealized environments, the gradients of heat flux, mixing depth, and surface roughness, as well as intense shear, updrafts, and subsidence, severely limit their suitability in coastal zones. During the past decade, mesoscale numerical models (MNMs) have gradually emerged as the basis of more advanced dispersion systems. The National Center for Atmospheric Research/Penn State MM4 model has successfully provided input to the Regional Acid Deposition Model (RADM). By the late 1980s, three-dimensional prognostic codes began to drive regional photochemical grid models, especially in California (Tesche, 1991).

The emergence of affordable high-speed computing portends increasingly widespread use of numerical codes in dispersion research. Many empirical models of the thermal internal boundary layer (TIBL) have been developed for coastal fumigation calculations (Venkatram, 1986). Stunder and Raman (1985) have tested and compared various TIBL formulations. Most do not address such factors as wind shear, variable initial air mass lapse rate, or coastal zone vertical motion. A high-resolution two-dimensional MNM provides a more generalized treatment of the TIBL. Three-dimensional MNMs account for such factors as heterogeneous land surfaces, complex topography, and irregular coastline shapes. By predicting the atmospheric state variables at each model grid point and time step, the MNM serves as the basis for new dispersion calculation methodologies.

New MNM-driven Lagrangian particle models, hybrid Lagrangian-Eulerian dispersion schemes, and source/receptor quantification methods show promise for the regulatory, operational, and research arenas (Uliasz and Pielke, 1990).

Considerable effort remains before new dispersion methodologies become generally applicable tools. Four-dimensional data assimilation techniques, successfully used in regional MNMs, must be adapted to the strong gradients of the coastal zone (Stauffer et al., 1990). The required intensive model evaluation studies will benefit from (1) improvements in technologies capable of resolving mesoscale vertical motion fields with a minimum detectable signal on the order of 10 cm/sec; (2) surface and airborne remote sensing systems that can map specific pollutant species (such as ozone); (3) improved sampling systems, including real-time airborne monitors, for longer-range (~100 km) tracers, such as perfluorocarbons for model evaluation and source attribution studies; and (4) a tracer that can be mapped using a volumetric remote sensing system over a relatively wide region. The interactions of coastal circulations with primary and photochemical aerosols and their treatment in regional meteorological and photochemical models are largely unaddressed issues in ozone and visibility modeling.

Few field programs have been designed to observe coastal circulations and mesoscale dispersion on both sides of the littoral with equal intensity and through their entire diurnal cycle. There is a dearth of comprehensive studies of dispersion in the land breeze. Improved coordination of large-scale air quality research efforts (which concentrate efforts over land) with offshore air-sea interaction studies would benefit both groups of investigators.

SUMMARY AND CONCLUSIONS

Transport and diffusion processes in coastal zones are only partially understood, and what knowledge there is, is weighted toward straight coastlines in flat terrain under idealized meteorological conditions. Dispersion modeling techniques currently in widespread use do not always account for even the known coastal zone dispersion mechanisms. Continued advances in affordable high-speed computing, prognostic mesoscale numerical modeling, and advanced dispersion simulation techniques promise to yield a more comprehensive understanding of coastal zone phenomena. Furthermore, these advanced technologies, once adequately tested, will soon be able to replace many current dispersion modeling systems used by the regulatory and emergency response communities.

The panel recommends the following:

• There should be continued development and testing of advanced prognostic meteorological and coupled dispersion models to simulate, and eventually operationally forecast, the impacts on pollutants of organized mesoscale vertical motions and intense horizontal and vertical gradients of wind, temperature, and turbulence found in three-dimensional, time-dependent coastal mesoscale regimes.

• Further investigations should be made of the role of coastal circulations in aggravating regional photochemical oxidant episodes, including interactions between locally emitted pollutants and those entering the region through long-range transport.

• Comprehensive tracer programs should be conducted at increasingly less idealized coastal sites, which would allow for evaluation, validation, and eventual widespread use of improved dispersion models, as well as further quantification of the complex transport and diffusion processes affecting pollutants in the coastal zone.

• Improved coordination should be required between air pollution and boundary layer field observation programs conducted on both sides of the littoral, which would advance knowledge of the underlying physical processes affecting dispersion over land, water, and the intervening transition zones.

8

Capabilities and Opportunities

OBSERVATIONAL TOOLS

Observations of coastal weather are required for routine as well as hazard forecasting, and also to improve our understanding of the many complex processes that have been described in previous chapters. Small space and time scales in the coastal zone place additional demands on observing systems compared to those used in more homogeneous environments. The difficulty and expense of installing observing systems in the coastal oceans bias observations toward land-based systems. Space-borne remote sensing has applications for the coastal zone, but again the coastal space and time scales are sometimes less than the scales resolved by satellite techniques. The ground-based, "local remote" sensors (radars, lidars, Doppler profilers) now being developed offer opportunities for increased measurements of coastal meteorology at appropriate scales. This chapter describes the state of present observations for operational coastal meteorology, techniques used in coastal research work, and future possibilities to improve both measurements and understanding.

In Situ Methods

The present operational coastal meteorological observations are obtained from a variety of systems operated primarily by the National Weather Service (NWS) of the National Oceanic and Atmospheric Administration (NOAA). Additional data are supplied by volunteer observing ships (VOSs). For the

FIGURE 8.1   Coastal rawinsonde observing network.

continental United States and the Great Lakes, NOAA operates 33 rawinsonde sites (see Figure 8.1) within 10 km of the coast. Rawinsonde observations are made twice a day at the synoptic standard times of 00 UTC and
12 UTC (0700 and 1900 EST), at these stations and others all over the
globe. A device with sensors for measuring pressure, temperature, and
relative humidity is carried aloft by a gas-filled balloon at about 300 m/min.
The device telemeters data back to a receiving station. Wind speed and
direction are calculated by tracking the motion of the balloon-borne instru-
ment. The data are normally processed into 6-sec averages of the desired
quantities. Observations at specified pressure levels are transmitted to weather
services worldwide. NOAA also operates, through the National Data Buoy
Center (NDBC), 26 instrumented buoys within 50 km of the coast or Great
Lakes shores (Figure 8.2). These are equipped with sensors for pressure,
wind, air temperature, and humidity at 3 or 10 m above the sea surface. In
addition, sea temperature and wave data are supplied. The buoys report
once per hour by satellite. Data retrieval rates are 90 percent for the meteorological sensors and 85 percent for the waves over a 5-year period. Some
vigilance by a broad-based user community, from fishing to environmental
research, may be necessary to keep buoys in place when their original mission has been served and the sponsor withdraws support. The VOSs report
data (pressure, temperatures, humidity, wind, and sea-state and cloud visual
observations) at bridge height (about 20 m for modern ships) every 6 hours
in the open ocean; within 200 km of land, NOAA requests reports every 3
hours. These disparate data—from different platforms at different times
and elevations—are blended into the overall data set by a computerized

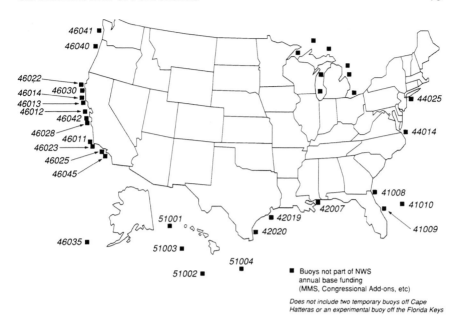

FIGURE 8.2   U.S. buoy network, September 30, 1991.

objective analysis scheme at NWS. For particular application to the coastal zone, we note that the standard synoptic observing times do not necessarily coincide with maxima or minima in the diurnal sea breeze cycle, and the effects of this bias may be more pronounced in the coastal zone than elsewhere; the lower levels of the balloon soundings do not often resolve the boundary layer and may reflect more of the local urban boundary layer than that over the nearby coast; there are no sounding data over the ocean; usual errors in reported ship positions are more critical than over the open ocean; and coastal ship data are naturally concentrated around ports.

For the future, NOAA's observational equipment modernization will offer some improvements and some degradation of data for the coastal zone. Of the 28 planned WSR-88 Doppler weather surveillance radars, 24 will provide over-water coverage. Reflectivity data (the return from precipitation water in clouds) will be available out to 300 to 400 km offshore; Doppler winds of the radial component will be available out to 150 km. These should result in an improved capability in offshore storm detection, movement, and development. The deployment locations, however, will not result in good coverage off the Pacific Northwest coast. Since there will also be returns from the sea surface as well for low antenna angles, there is the possibility that the "sea clutter" can be interpreted in terms of surface wind and wave data. NOAA's vertical Profiler Demonstration Network of

30 systems will be deployed only in the central and southern United States and therefore will not benefit the coastal zone. Only a few additional coastal NDBC buoy deployments are planned, and these are funded by agencies other than NOAA for specific periods as a part of ongoing experiments. No new upper-air rawinsonde stations are planned, and some coastal stations may be moved several tens of kilometers landward, eliminating real-time observations of the vertical thermodynamic structure within many coastal areas. One such site change, from Boothville to Slidell, Louisiana, has been commented on by Bosart (1990).

Coastal research experiments utilize existing synoptic data sources and, in addition, generally provide specialized observational systems of their own. These range from paying for additional NDBC buoy deployments; building and deploying separate weather buoys by oceanographic institutions (usually as the surface termination of an oceanographic mooring); deploying coastal weather stations to resolve mesoscale processes; using small upper-air balloon systems for soundings in particular locations; using special research aircraft equipped for meteorological measurements; taking meteorological measurements and soundings from ships; and using special satellite products. Coastal experiments usually cover the spatial range from less than 1 km to several hundred kilometers in order to resolve all of the meteorological processes; the experiments generally have an intensive phase of a month or two, sometimes imbedded in longer-term monitoring phases covering a season or more. For long-term coastal ocean studies over larger spatial scales, products such as the "Bakun" winds (Halliwell and Allen, 1984) are sometimes used, which are based on geostrophic winds.

Remote Sensing

Satellite remote sensing products are valuable for the coastal zone; the visual images from the GOES satellites show clearly the large-scale picture and often the sharp differences in cloud cover occurring right at the coast. Kelvin waves propagating up the west coast of the United States were tracked by Dorman (1985) from GOES images of the leading edge of the associated marine stratus cloud. High-quality photos from various manned space missions show dramatic modifications in cloud patterns near coastlines over the world. NOAA passive infrared imagery (e.g., AVHRR) interpreted as sea surface temperature often shows the overall characteristics of coastal oceanographic processes—cold water near the coast during wind-driven upwelling events (Kelly, 1985) and the "squirts and jets" of cold water often found off points and capes (Davis, 1985). For small-scale studies, data from the AVHRR sensors have to be regridded to achieve a spatial accuracy of 1 km. Similarly, the Coastal Zone Color Scanner (CZCS) provides high-resolution maps of plankton concentration. However, the spatial resolution of about

0.8 × 0.8 km is still more suited to the open ocean than the coastal zone. Pulse radars pointing at nadir are used to obtain the topography of the sea surface, from which geostrophic currents can be obtained. Synthetic aperture radar (SAR) has a fine spatial footprint (on the order of 100 m) and often shows complex signatures of the ocean surface, but possible interpretation of surface winds is a research project. Satellite microwave measurements (e.g., the Seasat scatterometer and the DMSP SSM/I) have spatial resolutions that are too coarse to be useful in studies of coastal meteorology.

At this time, the only active sensors in space are the altimeter, scatterometer, and SAR on the European Space Agency's Earth Resources Satellite ERS-1, and the SAR on an Almaz Soviet spacecraft. More are planned for the NASA/CNES TOPEX/POSEIDON mission in 1993, the U.S./Canadian RADARSAT, and the Japanese ADEOS launch in 1995. Passive sensors are on the NOAA series of satellites. The Sea-WIFS (Sea-Viewing Wide Field Sensor), scheduled for launch in 1993, will replace the CZCS with a more sensitive sensor.

Surface-based and space-based systems are complementary. The radar mentioned above is an obvious example of a surface-based system. Geographic coverage is less than from space-based systems, but spatial resolution is often better and thus of importance for the coastal zone. New technology applicable to coastal meteorology includes high-frequency, high-resolution atmospheric boundary layer (ABL) wind profilers; Doppler lidars, which are multichannel microwave radiometers for moisture and temperature profiles; and scanning and millimeter-wavelength radars. These can be based at the coast and pointed over land, over sea, or vertically to map mesoscale features in the coastal zone. In a coastal project Kropfli (1986) used a dual-Doppler radar system, which scattered off aircraft-deployed chaff in the Santa Barbara channel, to resolve an atmospheric eddy in the lee of Point Conception (see Figure 8.3).

The higher-frequency radars (915 MHz; Eklund et al., 1988) are able to backscatter continuously off turbulent refractive index fluctuations in the ABL and so are not restricted to clouds with precipitation or to the use of chaff. A 404-MHz vertically pointing system has been used in the Southern California bight to measure a strong nightly wind jet. With a colocated acoustic sounder to provide temperature profiles, these two systems will be able to map the time-height ABL structure at the coast with unprecedented resolution. Microwave radars have also been used to measure the liquid water content of coastal stratus clouds. Airborne or ground-based lidars can, in principle, provide information on wind, temperature, moisture, aerosols, and trace gases. A lidar was recently used on land in the Monterey, California, area to study the diurnal land-sea breeze; cross sections of the wind speed in Figure 8.4 show the classic land-sea breeze and also reveal complexities of the time evolution and spatial structure. Radar systems can

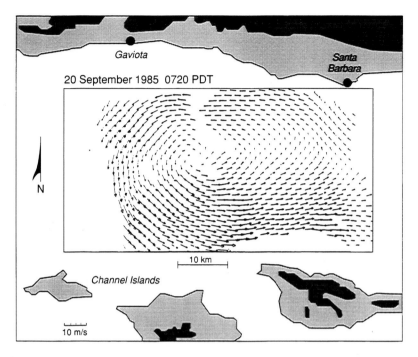

FIGURE 8.3   Dual-Doppler radar observations with the NOAA/WPL X-band systems in the Santa Barbara Channel using aircraft deployed chaff. A 1-hour composite of horizontal wind vectors in the lower atmospheric boundary layer shows the strong circulation (from Kropfli, 1986).

point at the sea surface and infer surface ocean current and perhaps wave information over mesoscale areas. These would be useful in combined oceanographic/meteorological studies of the coastal zone.

In summary, observations of the coastal zone are a special mix from different systems—land-based NWS synoptic reports, hourly data from coastal buoys, and VOS reports. The data are highly asymmetric; over the coastal ocean there are few surface observations and essentially no balloon soundings. The frequency of some of the observations may not be high enough to resolve important parts of coastal mesoscale processes. Satellite sensors have helped to identify some processes in the coastal zone, but quantitative measurements of important variables in addition to sea surface temperature at the appropriate scales are just beginning. The new WSR-88 radar systems will provide a dramatic increase of weather data off the coasts in storms. Newly developed surface-based remote sensors have the spatial resolution needed for coastal zone studies; they need to be deployed and tested in specialized research experiments to determine the seaward extent of their measurements. Coastal experiments will undoubtedly continue to deploy additional specialized buoys, sounding systems, research aircraft, ships, and the newer remote systems as they become available.

Two measurement scales appear to be particularly important. Long temporal and large spatial scales provide the framework for more detailed mesoscale process-oriented studies. Low-frequency measurements would depend on buoys capable of measuring atmospheric and oceanographic parameters. Smaller-scale studies would depend on an extensive array of "air-

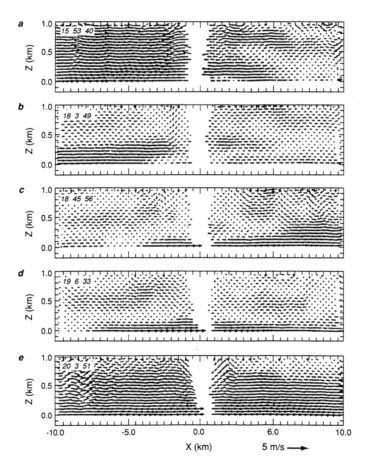

FIGURE 8.4 Doppler lidar wind vector measurements in a plane perpendicular to the coast near Monterey, California, depicting a 4-hour period in the transition from a land breeze to a sea breeze (sunrise is about 1400 UTC). The lidar is located at (0,0), 1.5 km inland; Monterey Bay is to the left; and land is to the right (from Olivier et al., 1991).

sea interaction" buoys and moorings, land-based meteorological sites, and in situ measurements with ships and aircraft. The buoys should be capable of making accurate and repeated measurements of sea surface temperature, air temperature, humidity, barometric pressure, wind velocity, solar and long-wave radiation, precipitation, and trace gas concentrations from which fluxes can be estimated. In addition, current and surface wave measurements would be required. Ideally, direct covariance estimates of surface fluxes are needed; however, these remain beyond the reach of present technology for long-term measurements on unmanned platforms.

Reliable, remotely operated instrumentation for atmospheric measurements continues to be a challenge. For example, the ability to determine the depth of the boundary layer and cloud cover from a buoy would provide a significant amount of information about the structure of the atmosphere and the response of the ocean. Understanding the interaction between the atmosphere and the ocean requires an understanding of the structure and the vertical and horizontal variability of the entire atmospheric and oceanic boundary layers. Hitherto, this kind of detailed information has not been available for coastal studies; it would provide a substantial contribution to our understanding of the mesoscale coastal ocean-atmosphere system.

MODELING TECHNOLOGY

Modeling of coastal processes requires the application of physical and numerical simulation tools and analytical studies. The use of physical models related to coastal studies has focused on roughness and heat changes in the near-shore environment as air moves from ocean to land and vice versa. Avissar et al. (1990) have summarized the spatial scales that can be simulated accurately by wind tunnel modeling. One conclusion is that laboratory model studies in coastal environments are inappropriate when the influence of the earth's rotation becomes important. Also, it was shown that coupling meteorological wind tunnel and mesoscale numerical models offers a larger range of conditions than can be simulated by either approach alone. For example, a numerical simulation of a sea breeze can be used to provide the scaled winds and turbulence for input to a detailed wind tunnel simulation of flow around buildings in the coastal environment.

Numerical modeling and analytical modeling have been applied more generally to coastal meteorological studies. Pielke (1984, 1989) reviewed a range of models that have been developed to study mesoscale processes, including land-water interfaces. Zanetti (1990) has summarized air pollution models, including those that are applicable to coastal environments, as well as deficiencies in current regulatory models in this environment. Anthes et al. (1982) review regional models that necessarily include the ocean-continent interface within these simulation tools. To represent the structure of the coastal environment adequately, high spatial resolution is required. Lyons et al. (1987), for example, have demonstrated for the Kennedy Space Center environment that horizontal grid increments larger than 1 km are unable to realistically simulate strong vertical motions associated with the Indian and Banana rivers. Analytical modeling has helped to clarify understanding of coastal flows. Rotunno (1983) and Dalu and Pielke (1989), for example, have evaluated theoretically the influence of latitude on the propagation characteristics of the sea breeze.

Continued application of these modeling tools is desirable for continued improvement in our understanding of coastal flows and in developing predictive capabilities. Predictability can be defined as the limit in time to which a prediction has skill, as measured, for example, against climatology or persistence. In the context of nonlinear dynamics for a deterministic dissipative dynamic system such as the atmosphere, predictability is limited by the sensitivity to initial conditions (Lorenz, 1963). The sea breeze is an example of a surface-forced coastal feature that results from differential heating of land and water. A coherent wind circulation is due to this well-defined surface heating pattern. One question that needs to be addressed is: How does the predictability of this mesoscale feature change as a function of the spatial size of the differential heating regions, the depth of the convective mixed layer, etc.? A second question is: To what extent are small-scale meteorological features, such as divergence in the wind field phase, locked to coastal topography such as capes in a variety of transitory land-falling storms (e.g., Zemba and Friehe, 1987; Mass and Ferber, 1990)? This "sensitivity to boundary" condition needs to be investigated in the context of nonlinear dynamics.

The emergence of high-performance computer workstations having substantial fractions of the calculation speed and superior throughput of present-day mainframe supercomputers is having an effect on mesoscale research and modeling. These computers allow researchers to run mesoscale models at fine mesh sizes (i.e., time steps) and conduct numerous sensitivity studies. They can also provide effective visualization capabilities and hard disk data storage. These workstations will permit central supercomputers to concentrate on the modeling of the coastal zone that cannot be accomplished at workstations. Recent work (Lyons et al., 1991b) has revealed that sea breeze simulations using nested grid nonhydrostatic models employing very fine inner nests (1000 m or less) produce much sharper frontal discontinuities and consequently stronger vertical motions than do previous coarser-mesh runs (see Figure 3.1). When complex shorelines and topography are involved, the resulting simulations become both quantitatively and qualitatively different. Substantial model sensitivity tests need to be conducted to assess how such factors as horizontal and vertical mesh size, as well as characterization of surface heat and moisture fluxes, influence the results.

Data such as that collected during the 1991 Convection and Precipitation/Electrification (CAPE) program will be among the most complete for model sensitivity testing and evaluation. Some present views of coastal mesoscale processes may require revision in light of very fine mesh simulations. The issue of just how fine is fine enough remains open.

Four-dimensional data assimilation (4DDA) issues also need to be addressed. While 4DDA has been used with success at meso-alpha and regional scales, the use of data in coastal zones, where extreme vertical and horizontal gradients are common, raises difficult issues of spatial and temporal representativeness of the data to be used for 4DDA. Pielke et al. (1989) demonstrate the need for increased accuracy in measured data as their spatial separation decreases, showing a general inability to relocate time-space cross sections in the coastal environment. Any new observational system installed in coastal zones, if it is to be useful for potential 4DDA applications, must be carefully designed. If operational coastal models are to become a reality, frequently updated regional soil moisture and vegetation information (from observations not now readily available) will become a requirement. Four-dimensional data assimilation techniques currently being investigated include those typified by Errico and Vukicevic (1991), Hoke and Anthes (1976), and Lipton and Pielke (1986).

The following specific recommendations are made:

• Site-specific three-dimensional numerical modeling of coastal zones and studies of coastal zone physics may be enhanced by onsite use of high-performance workstations. The development of massively parallel workstations and user-friendly software to increase this performance capability should be encouraged.

• Standardized procedures to construct subroutines should be developed in order to facilitate the exchange of model components between investigators.

• The value of ensemble model predictions, as contrasted with a collection of model realizations, needs to be determined. Such evaluations should include the testing of subgrid-scale parameterizations that incorporate a stochastic component in order to ascertain the significance of deviations from an ensemble representation.

• The extent to which predictability is extended as a result of the spatial structure of the surface forcing needs to be established.

• Requirements to adequately utilize 4DDA techniques in the coastal environment need to be assessed. Adjoint methods, nudging procedures, and variants of normal model analyses should be tested in the coastal environment where differences in land and water forcing could perhaps be applied to improve the accuracy of data assimilation.

9

Educational and Human Resources

Compared to other subdisciplines in meteorology, many areas of coastal meteorology lack focus and activity. This is reflected in the general dearth of basic research related to coastal meteorology. An illustration of this point can be made. The American Meteorological Society's (AMS) Scientific and Technological Activities Commission maintains a Committee on Meteorology and Oceanography of the Coastal Zone. In May 1991, the AMS and this committee sponsored the Fifth Conference on Meteorology and Oceanography of the Coastal Zone, in Miami, Florida. Forty papers were presented at the conference from a total of 74 contributing authors and coauthors. The emphasis of the papers was on applied and operational meteorology (e.g., storm surges), rather than substantive theoretical problems of the field. The small showing for this conference and the narrow range of interest represented by the contributions reflect the general lack of basic research in the area of coastal meteorology.

Other than a general lack of activity in coastal meteorology, the manpower issues that face coastal meteorology are the same as those facing the larger disciplines of meteorology and oceanography, and these will not be detailed in this report. Here the focus will be on two issues related to education: (1) the breadth of education required to address many of the problems of coastal meteorology and (2) the need for improved education in instrumentation and observational systems.

To address many of the outstanding problems in coastal meteorology as outlined in this report, a broadly based curriculum in meteorology and the supporting sciences is needed. Interdisciplinary problems, such as air-sea interaction and air pollution in the coastal zone, require education in ocean-ography, chemistry, and biology, in addition to graduate-level courses in both atmospheric physics and dynamics.

Because of the substantial need for improvements in the application of sophisticated observational tools to the problems of coastal meteorology (see section above), the recent report by the committee for a Study on Observational Systems (SOS), which was formed jointly by the AMS and the University Corporation for Atmospheric Research, is especially relevant. The SOS committee was organized to address the concern that development of instrumentation measurement techniques is outpacing the training and education of those who will use them. The committee concluded (SOS, 1991) that "there is a widening gap between the sophistication and complexity of the state of the art in measurement technology and [the] abilities of universities to offer relevant and adequate curricula." Universities are finding it increasingly difficult to maintain even a basic instructional program in observational systems and attendant laboratory facilities. The following recommendations made by the SOS committee and adopted here are of particular relevance to coastal meteorology:

• Atmospheric science departments should provide more comprehensive curricula on observational systems and experimental methods and integrate these with curricula that focus on the theory of atmospheric and oceanic processes.

• Universities should acquire modern observational equipment to provide hands-on experience to students. The federal funding agencies should assist in the provision of these facilities.

• National centers and laboratories should devise programs that encourage scientists from universities and other agencies to participate in collaborative activities that broaden exposure to and share expertise in modern observational systems.

Other panel recommendations for generating greater interest in advancing understanding and application of coastal meteorology are as follows:

• There should be further delineation of the field through such means as this report; AMS, American Geophysical Union, and Oceanographic Society committees; and relevant agencies, such as the National Science Foundation's Committee on the Coastal Ocean.

• Conferences and symposia should be organized on the subject of basic scientific issues in coastal meteorology and oceanography to encourage collaboration between meteorologists, oceanographers, and other specialists in such fields as atmospheric and marine chemistry and remote sensing.

• Short courses or workshops should be conducted on coastal meteorology, patterned after either the AMS short courses or the National Center for

Atmospheric Research workshops. Such activities should result in publication of monographs, textbooks, or laboratory workbooks for wider classroom applications.

• Specific courses related to the meteorology of the coastal zone should be offered in college curricula.

References

Ackerman, B.W. 1981. Mesoscale Atmospheric Circulations. Academic Press, London. 495 pp.

André, J.-C., P. Bougeault, and J.-P. Goutorbe. 1990. Regional estimates of heat and evaporation fluxes over non-homogeneous terrain. Examples from the HAPEX-MOBILHY Programme. Bound. Layer Meteorol. 50:77-108.

Anthes, R.A., Y.-H. Kuo, S.G. Benjamin, and Y.-F. Li. 1982. The evolution of the mesoscale environment of severe local storms: Preliminary modeling results. Mon. Weather Rev. 110:1187-1213.

Avissar, R., M.D. Moran, R.A. Pielke, G. Wu, and R.N. Meroney. 1990. Operating ranges of mesoscale numerical models and meteorological wind tunnels for the simulation of sea and land breezes. Bound. Layer Meteorol. Special Anniversary Issue, Golden Jubilee 50:227-275.

Baes, C.F., and G.G. Killough. 1985. A two-dimensional CO_2-ocean model including the biological processes. U.S. Department of Energy, Report DoE/NBB-0070 TR021, Washington, D.C.

Bane, Jr., J.M., and K.E. Osgood. 1989. Winter-time air-sea interaction processes across the Gulf Stream. J. Geophys. Res. 94:10755-10772.

Bane, J.M., C.D. Winant, and J.E. Overland. 1990. Planning for coastal air-sea interaction studies in COPO. Bull. Am. Meteorol. Soc. 71:514-519.

Bannon, P.R. 1981. Synoptic-scale forcing of coastal lows. Forced double Kelvin waves in the atmosphere. Q. J. R. Meteorol. Soc. 107:313-327.

Baralt, G.L., and R.A. Brown. 1965. The Land and Sea Breeze: An Annotated Bibliography. Article F, Final Report, Mesometeorological Field Studies, Department of Geophysical Sciences, University of Chicago. 61 pp.

Barry, R.G. 1986. Meteorology of the seasonal sea ice zone. Pp. 993-1020 in Geophysics of Sea Ice. N. Untersteiner (ed.). Plenum, New York.

Bates, T.S., R.J. Charlson, and R.H. Gammon. 1987. Evidence for the climatic role of marine biogenic sulfur. Nature 329:319-321.

Beardsley, R.C., C.E. Dorman, C.A. Friehe, L.K. Rosenfeld, and C.D. Winant. 1987. Local atmospheric forcing during the coastal ocean dynamics experiment. 1. A description of the marine boundary layer and atmospheric conditions over a northern California upwelling region. J. Geophys. Res. 92:1467-1488.

Beljaars, A.C.M., and A.A.M. Holtslag. 1991. Flux parameterization over land surfaces for atmospheric models. J. Atmos. Sci. 30:327-341.

Bell, G.D., and L.F. Bosart. 1988. Appalachian cold-air damming. Mon. Weather Rev. 116:137-161.

Bender, M.A., R.E. Tuleya, and Y. Kurihara. 1987. A numerical study of the effect of island terrain on tropical cyclones. Mon. Weather Rev. 115:130-155.

Bennett, T.J., and K. Hunkins. 1986. Atmospheric boundary layer modification in the marginal ice zone. J. Geophys. Res. 91:13033-13044.

Binschadler, R.A. (ed.). 1991. West Antarctic Ice Sheet Initiative Science and Implementation Plan. Proceedings of NASA Conference held at NASA Goddard Space Flight Center, Greenbelt, Md., Oct. 16-18, 1990. 51 pp.

Biswas, K.R., and P.V. Hobbs. 1990. Lightning over the Gulfstream. Geophys. Res. Lett. 17:941-943.

Bond, N.A., and S.A. Macklin. 1992. Aircraft observations of offshore directed flow near Wide Bay, Alaska. Mon. Weather Rev. (in review).

Bond, N.A., and M.A. Shapiro. 1991. Polar lows over the Gulf of Alaska in conditions of reverse shear. Mon. Weather Rev. 119:551-572.

Bosart, L.F. 1975. New England coastal frontogenesis. Q. J. R. Meteorol. Soc. 101:957-978.

Bosart, L.F. 1981. The Presidents' Day snowstorm of 18-19 February 1979: A subsynoptic-scale event. Mon. Weather Rev. 109:1542-1566.

Bosart, L.F. 1983. Analysis of a California Catalina eddy event. Mon. Weather Rev. 111:1619-1633.

Bosart, L.F. 1984. The Texas coastal rainstorm of 17-21 September 1979: An example of synoptic-mesoscale interaction. Mon. Weather Rev. 112:1108-1133.

Bosart, L.F. 1990. Degradation of the North American rawinsonde network. Weather Forecast. 5:689-690.

Bosart, L.F., C.J. Vaudo, and J.H. Helsdon, Jr. 1972. Coastal frontogenesis. J. Appl. Meteorol. 11:1236-1258.

Bosart, L.F., V. Pagnotti, and B. Lettau. 1973. Climatological aspects of eastern United States back-door cold frontal passages. Mon. Weather Rev. 101:627-635.

Broecker, W.S. 1982. Ocean chemistry during glacial time. Geochim. Cosmochim. Acta 46:1689-1705.

Broecker, W.S., and T.H. Peng. 1974. Gas exchange rates between air and sea. Tellus 26:21-35.

Broecker, W.S., D.M. Peteet, and D. Rind. 1985. Does the ocean-atmosphere system have more than one stable mode of operation? Nature 315:21-26.

Bromwich, D.H. 1988. Snowfall in high southern latitudes. Rev. Geophys. 26:149-168.

Bromwich, D.H. 1989. An extraordinary katabatic wind regime at Terra Nova Bay, Antarctica. Mon. Weather Rev. 117:688-695.

Bromwich, D.H. 1991. Mesoscale cyclogenesis over the southwestern Ross Sea linked to strong katabatic winds. Mon. Weather Rev. 119:1736-1752.

Bromwich, D.H., and D.D. Kurtz. 1984. Katabatic wind forcing of the Terra Nova Bay polynya. J. Geophys. Res. 89:3561-3572.

Brost, R.A., J.C. Wyngaard, and D.H. Lenschow. 1982. Marine stratocumulus layers. Part II: Turbulence budgets. J. Atmos. Sci. 39:818-836.

Burpee, R.W., and L.N. Lahiff. 1984. Area averaged rainfall variations on sea breeze days in South Florida. Mon. Weather Rev. 112:520-534.

Businger, J.A. 1986. Evaluation of the accuracy with which dry deposition can be measured with current micrometeorological techniques. J. Clim. Appl. Meteorol. 25:1100-1124.

Businger, J.A., and A.C. Delany. 1990. Chemical sensor resolution required for measuring surface fluxes by three common micrometeorological techniques. J. Atmos. Chem. 10:399-410.

Businger, J.A., and S.P. Oncley. 1990. Flux measurements with conditional sampling. J. Atmos. Ocean. Tech. 7:349-352.

Businger, S. 1991. Arctic hurricanes. Am. Sci. 79:18-33.

Businger, S., and B. Walter. 1988. Comma cloud development and associated rapid cyclogenesis over the Gulf of Alaska: A case study using aircraft and operational data. Mon. Weather Rev. 116:1103-1123.

Businger, S., and R.J. Reed. 1989. Cyclogenesis in cold air masses. Weather Forecast. 2:133-156.

Chandik, J.F., and W.A. Lyons. 1971. Thunderstorms and the lake breeze front. Pp. 218-225 in Proceedings, Seventh Conference on Severe Local Storms, Kansas City, Mo. American Meteorological Society, Boston.

Charlson, R.J., J.E. Lovelock, M.O. Andreae, and S.G. Warren. 1987. Oceanic phytoplankton, atmospheric sulphur, cloud albedo and climate. Nature 326:655-661.

Charnock, H. 1979. Air-sea interaction. In Evolution of Physical Oceanography. B.A. Warren and C. Wunsch (eds.). The MIT Press, Cambridge, Mass. 623 pp.

Charnock, H., and J.A. Businger. 1991. The frontal air-sea interaction experiment in perspective. J. Geophys. Res. 96:8639-8642.

Chen, W.D., and R.B. Smith. 1987. Blocking and deflection of airflow by the Alps. Mon. Weather Rev. 115:2578-2597.

Chen, G.T.J., and C.C. Yu. 1988. Study of low-level jet and extremely heavy rainfall over northern Taiwan in the Mei-Yu season. Mon. Weather Rev. 116:884-891.

Clancy, R.M., J.D. Thompson, H.E. Hulbert, and J.D. Lee. 1979. A model of mesoscale air-sea interaction in a sea-breeze coastal upwelling regime. Mon. Weather Rev. 107:1478-1505.

Clark, J.H.E., and S.R. Dembek. 1991. The Catalina eddy event of July 1987: A coastally trapped mesoscale response to synoptic forcing. Mon. Weather Rev. 119:1714-1735.

Claussen, M. 1987. The flow in a turbulent boundary layer upstream of a change in surface roughness. Bound. Layer Meteorol. 40:31-86.

Claussen, M. 1991. Estimation of areally averaged surface fluxes. Bound. Layer Meteorol. 54:387-410.

Colquhoun, J.R., D.J. Shepherd, C.E. Coulman, R.K. Smith, and K. McInnes. 1985. The southerly buster of southeastern Australia: An orographically forced cold front. Mon. Weather Rev. 113:2090-2107.

Crane, A.J. 1988. The use of tracers in modeling the oceanic uptake of carbon dioxide. In Tracers in the Ocean. Proceedings of a Royal Society Discussion Meeting. H. Charnock, J.E. Lovelock, P.S. Liss, and M. Whitfield (eds.). Royal Society, London.

Curry, J.A. 1983. On the formation of continental polar air. J. Atmos. Sci. 40:2278-2292.

Curry, J.A., F.G. Meyer, L.F. Radke, C.A. Brock, and E.E. Ebert. 1990. Occurrence and characteristics of lower tropospheric ice crystals in the Arctic. Int. J. Climatol. 10:749-764.

Dalu, G.A., and R.A. Pielke. 1989. An analytical study of the sea breeze. J. Atmos. Sci. 46:1815-1825.

Davies, H.C., and H. Pichler. 1990. Mountain meteorology and ALPEX—An introduction. Meteorol. Atmos. Phys. 43:3-4.

Davis, R.E. 1985. Drifter observations of coastal surface currents during CODE: The method and descriptive view. J. Geophys. Res. 90:4741-4755.

Davis, R.E., and P.S. Bogden. 1989. Variability on the California shelf forced by local and remote winds during the coastal ocean dynamics experiment. J. Geophys. Res. 94:4763-4787.

Dawe, A.J. 1982. A study of katabatic wind at Brugge on 27 February 1975. Meteorol. Mag. 111:1-13.

Deardorff, J.W. 1974. Three-dimensional numerical study of turbulence in an entraining mixed layer. Bound. Layer Meteorol. 7:199-226.

Deardorff, J.W. 1980. Stratocumulus-capped mixed layers derived from a three-dimensional model. Bound. Layer Meteorol. 18:495-527.

Deardorff, J.W., and G.E. Willis. 1982. Dependence of mixed-layer entrainment on shear stress and velocity jump. J. Fluid Mech. 115:123-149.

Defant, F. 1950. Theorie der Land- und Seewinde. Archiv für Meteorologie, Geophysik, und Bioklimatologie. Vienna. Ser. A, 2-3:404-425.

Defant, F. 1951. Local winds. Pp. 655-672 in Compendium of Meteorology. T.F. Malone (ed.). American Meteorological Society, Boston.

Dempsey, D.P., and R. Rotunno. 1988. Topographic generation of mesoscale vortices in mixed-layer models. J. Atmos. Sci. 45:2961-2978.

Denbo, D.W., and J.S. Allen. 1987. Large-scale response to atmospheric forcing of shelf currents during CODE: The statistical and dynamical views. J. Geophys. Res. 92:1757-1782.

Donelan, M.A., J. Hamilton, and W.H. Hui. 1985. Directional wave spectra of wind generated waves. Philos. Trans. R. Soc. London, Ser. A, 315:509-562.

Dorman, C.E. 1985. Evidence of Kelvin waves in California's marine layer and related eddy generation. Mon. Weather Rev. 113:827-839.

Dorman, C.E. 1987. Possible role of gravity currents in northern California's coastal summer wind reversals. J. Geophys. Res. 92:1497-1506.

Draghici, I. 1984. Black Sea coastal frontogenesis. Pp. 75-79 in Nowcasting II:

Mesoscale Observations and Very Short-Range Weather Forecasting. European Space Agency, Paris, France.

Eklund, W.L., D.A. Carter, and B.B. Balsley. 1988. A UHF wind profiler for the boundary layer: Brief description and initial results. J. Atmos. Ocean Tech. 5:432-441.

Elliot, D.L., and J.J. O'Brien. 1977. Observational studies of the marine boundary layer over an upwelling region. Mon. Weather Rev. 105:86-98.

Emanuel, K., and R. Rotunno. 1989. Polar lows as arctic hurricanes. Tellus 41A:1-17.

Enriquez, A., and C.A. Friehe. 1991. Variability of the atmospheric boundary layer over a coastal shelf in winter during SMILE. Pp. 102-107 in Fifth Conference on Meteorology and Oceanography of the Coastal Zone. American Meteorological Society, Boston.

Errico, R.M., and T. Vukicevic. 1991. Sensitivity analysis using an adjoint of the PSU/NCAR mesoscale model. National Center for Atmospheric Research, #0501/91-3, Boulder, Colo.

Fairall, C.W., and S.E. Larsen. 1986. Inertial-dissipation methods and turbulence fluxes at the air-ocean interface. Bound. Layer Meteorol. 11:19-38.

Fairall, C.W., K.L. Davidson, and G.E. Schacher. 1982. Meteorological models for optical properties in the marine atmospheric boundary layer. Opt. Eng. 21:847-857.

Ferber, G.K., and C.F. Mass. 1990. Surface pressure perturbations produced by an isolated mesoscale topographic barrier. Part II: Influence on regional circulation. Mon. Weather Rev. 118:2597-2606.

Fitzjarrald, D.R. 1986. Slope winds in Veracruz. J. Clim. Appl. Meteorol. 25:133-144.

Forbes, G.S., R.A. Anthes, and D.W. Thomson. 1987. Synoptic and mesoscale aspects of an Appalachian ice storm associated with cold-air damming. Mon. Weather Rev. 115:564-591.

Friehe, C.A., W.J. Shaw, D.P. Rogers, K.L. Davidson, W.G. Large, S.A. Stage, G.H. Crescenti, S.J.S. Khalsa, G.K. Greenhut, and F. Li. 1991. Air-sea fluxes and surface-layer turbulence around a sea surface temperature front. J. Geophys. Res. 96:8593-8609.

Fujibe, F. 1990. Climatology of the coastal front in the Kanto Plain. Papers Meteorol. Geophys. 41:105-128.

Garratt, J.R. 1987. The stably stratified internal boundary layer for steady and diurnally varying offshore flow. Bound. Layer Meteorol. 38:369-394.

Geernaert, G. 1988. Drag coefficient modeling for the near coastal zone. Dynam. Atmos. Oceans 11:307-322.

Geernaert, G.L. 1990. Bulk parameterizations for the wind stress and heat fluxes. Pp. 91-172 in Surface Waves and Fluxes: Theory and Remote Sensing. Vol. 1: Current Theory. G.L. Geernaert and W.J. Plant (eds.). Kluwer Academic Publisher, Norwell, Mass. 336 pp.

Geernaert, G.L., K.B. Katsaros, and K. Richter. 1986. Variation of the drag coefficient and its dependence on sea state. J. Geophys. Res. 91:7667-7679.

Geernaert, G.L., S.E. Larsen, and F. Hansen. 1987. Measurements of the wind stress, heat flux, and turbulence intensity during storm conditions over the North Sea. J. Geophys. Res. 92:13127-13139.

Gill, A.E. 1977. Coastally trapped waves in the atmosphere. Q. J. R. Meteorol. Soc. 103:431-440.

Gossard, E., and W. Munk. 1954. On gravity waves in the atmosphere. J. Meteorol. 11:259-269.

Gray, J., and J.E. Overland. 1986. Meteorology. Pp. 31-56 in Gulf of Alaska: Physical Environment and Biological Resources. D.W. Hood and S.T. Zimmerman (eds.). U.S. Government Printing Office, Washington, D.C.

Gryning, S.E., and E. Lyck. 1983. A tracer investigation of the atmospheric dispersion in the Dyrnaes Valley, Greenland. Risø National Laboratory, Denmark, Report Risø-R-481.

Gutman, L.N. 1983. On the theory of the katabatic slope wind. Tellus 35A:213-218.

Hadfield, M.G., W.R. Cotton, and R.A. Pielke. 1991. Large-eddy simulations of thermally forced circulations in the convective boundary layer. Part I: A small-scale circulation with zero wind. Bound. Layer Meteorol. 57:79-114.

Hadfield, M.G., W.R. Cotton, and R.A. Pielke. 1992. Large-eddy simulations of thermally forced circulations in the convective boundary layer. Part II: The effect of changes in wavelength and wind speed. Bound. Layer Meteorol. 58:307-328.

Halliwell, Jr., G.R., and J.S. Allen. 1984. Large-scale sea level response to atmospheric forcing along the west coast of North America, summer 1973. J. Phys. Ocean. 14:864-886.

Halliwell, Jr., G.R., and J.S. Allen. 1987. The large-scale coastal wind field along the west coast of North America. J. Geophys. Res. 92:1861-1884.

Haurwitz, B. 1947. Comments on the sea-breeze circulation. J. Meteorol. 4:1-8.

Hegg, D.A., R.J. Ferek, P.V. Hobbs, and L.F. Radke. 1991. Dimethyl sulfide and cloud condensation nucleus concentrations in the Northeast Pacific Ocean. J. Geophys. Res. 96:13189-13191.

Herman, A.J., B.M. Hickey, C.F. Mass, and M.D. Albright. 1990. Orographically trapped coastal wind events in the Pacific Northwest and their ocean response. J. Geophys. Res. 95:13169-13193.

Hibler, III, W.D. 1979. A dynamic thermodynamic sea ice model. J. Phys. Oceanogr. 9:815-846.

Hignett, P. 1991. Observations of diurnal variation in a cloud-capped marine boundary layer. J. Atmos. Sci. 48:1474-1482.

Hobbs, P.V. 1987. The Gulfstream rainband. Geophys. Res. Lett. 14:1142-1145.

Hobbs, P.V., T.J. Matejka, P.H. Herzegh, J.D. Locatelli, and R.A. Houze, Jr. 1980. The mesoscale and microscale structure and organization of clouds and precipitation in midlatitude cyclones. I. A case study of a cold front. J. Atmos. Sci. 37:586-596.

Hoke, J.E., and R.A. Anthes. 1976. The initialization of numerical models by a dynamic initialization technique. Mon. Weather Rev. 104:1551-1556.

Holland, G.J., and L.M. Leslie. 1986. Ducted coastal ridging over S.E. Australia. Q. J. R. Meteorol. Soc. 112:731-748.

Holt, T., and S. Raman. 1988. A review and comparative evaluation of multilevel boundary layer parameterizations for first-order and turbulent kinetic energy closure schemes. Rev. Geophys. 26:761-780.

Holt, T., and S. Raman. 1990. Marine boundary layer structure and circulation in the region of offshore redevelopment of a cyclone during GALE. Mon. Weather Rev. 118:392-410.

Howells, P.A.C., and Y.-H. Kuo. 1988. A numerical study of the mesoscale environment of a southerly buster event. Mon. Weather Rev. 116:1171-1178.

Huang, N.E., L.F. Bliven, S.R. Long, and P.S. DeLeonibus. 1986. A study of the relationship among wind speed, sea state, and the drag coefficient for a developing wave field. J. Geophys. Res. 91:7733-7742.

Jehn, K.H. 1973. A Sea Breeze Bibliography, 1664-1972. Atmospheric Sciences Group, University of Texas, Austin, Report No. 37. 51 pp.

Johannessen, O.M., J.A. Johannessen, and S. Sandven. 1988. Mesoscale ocean processes in the marginal ice zone. EOS 69:1276.

Kaimal, J.C., J.C. Wyngaard, D.A. Haugen, O.R. Cote, Y. Izumi, S.J. Caughey, and C.J. Readings. 1976. Turbulence structure in the convective boundary layer. J. Atmos. Sci. 33:2152-2169.

Kelly, K.A. 1985. The influence of winds and topography on the sea surface temperature patterns over the northern California slope. J. Geophys. Res. 90:11783-11798.

Keshishian, L.G., and L.F. Bosart. 1987. A case study of extended East Coast frontogenesis. Mon. Weather Rev. 115:100-117.

Kropfli, R.A. 1986. Airflow characteristics in the marine boundary layer over the Santa Barbara Channel during SCCCAMP: Part I. Summary of dual-Doppler radar observations. NOAA Technical Memorandum ERL WPL-141, Boulder, Colo. 105 pp.

Kuo, Y.-H., R.J. Reed, and S. Low-Nam. 1991. Effects of surface energy fluxes during the early development and rapid intensification stages of seven explosive cyclones in the western Atlantic. Mon. Weather Rev. 119:457-476.

Lackmann, G.M., and J.E. Overland. 1989. Atmospheric structure and momentum balance during a gap-wind event in Shelikof Strait, Alaska. Mon. Weather Rev. 117:1818-1833.

Lapenta, W.M., and N.L. Seaman. 1990. A numerical investigation of East Coast cyclogenesis during the cold-air damming event of 27-28 February 1982. Part I: Dynamic and thermodynamic structure. Mon. Weather Rev. 118:2668-2695.

Lee, T.N., and L.P. Atkinson. 1983. Low-frequency current and temperature variability from Gulf Stream frontal eddies and atmospheric forcing along the southeast U.S. outer continental shelf. J. Geophys. Res. 88:4541-4567.

Lee, T.N., E. Williams, J. Wang, and R. Evans. 1989. Response of the South Carolina continental shelf waters to wind and Gulf Stream forcing during winter of 1986. J. Geophys. Res. 94:10715-10754.

Lenschow, D.H. 1973. Two examples of planetary boundary layer modification over the Great Lakes. J. Atmos. Sci. 30:586-581.

Lenschow, D.H., J.C. Wyngaard, and W.T. Pennell. 1980. Mean-field and second-moment budgets in a baroclinic, convective boundary layer. J. Atmos. Sci. 37:1313-1326.

Lettau, H.H., and W. Schwerdtfeger. 1967. Dynamics of the surface wind regime over the interior of Antarctica. Antarctic Journal of the U.S. 2:155-158.

Lipton, A.E., and R.A. Pielke. 1986. Vertical normal modes of a mesoscale model using a scaled height coordinate. J. Atmos. Sci. 43:1650-1655.

Liss, P.S. 1988. Tracers of air-sea gas exchange. In Tracers in the Ocean. Proceedings of a Royal Society Discussion Meeting. H. Charnock, J.E. Lovelock, P.S. Liss, and M. Whitfield (eds.). Royal Society, London.

Liss, P.S., and L. Merlivat. 1986. Air-sea gas exchange rates: Introduction and synthesis. Pp. 113-127 in The Role of Air-Sea Exchange in Geochemical Cycling. Adv. Sci. Inst. Ser., P. Buat-Menard (ed.). D. Reidel, Norwell, Mass.

Lopez, M.E., and W.E. Howell. 1967. Katabatic winds in the equatorial Andes. J. Atmos. Sci. 24:29-35.

Lorenz, E. 1963. Deterministic nonperiodic flow. J. Atmos. Sci. 20:130-141.

Lyons, W.A. 1972. The climatology and prediction of the Chicago lake breeze. J. Appl. Meteorol. 11:1259-1270.

Lyons, W.A. 1975. Turbulent diffusion and pollutant transport in shoreline environments. Pp. 135-208 in Lectures on Air Pollution and Environmental Impact Analysis. D.A. Haugen (ed.). American Meteorological Society, Boston.

Lyons, W.A., and H.S. Cole. 1976. Photochemical oxidant transport: Mesoscale lake breeze and synoptic-scale aspects. J. Appl. Meteorol. 15:733-744.

Lyons, W.A., C.S. Keen, and J.A. Schuh. 1983. Modeling Mesoscale Diffusion and Transport Processes for Releases Within Coastal Zones During Land/Sea Breezes. U.S. Nuclear Regulatory Commission, NUREG/CR-3542, Washington, D.C. 183 pp.

Lyons, W.A., J.A. Schuh, D. Moon, R.A. Pielke, W.R. Cotton, and R.W. Arritt. 1987. Short-range forecasting of sea breeze generated thunderstorms at the Kennedy Space Center: A real-time experiment using a primitive equation mesoscale numerical model. Pp. 503-508 in Proceedings of the Symposium on Mesoscale Analysis and Forecasting, Incorporating Nowcasting, August 17-19, Vancouver, British Columbia, Canada.

Lyons, W.A., R.A. Pielke, W.R. Cotton, C.J. Tremback, R.L. Walko, and C.S. Keen. 1991a. The impact of wind discontinuities and associated zones of strong vertical motions upon the mesoscale dispersion of toxic airborne contaminants in coastal zones. Proceedings, Joint Army-Navy-NASA-Air Force Safety and Environmental Protection Subcommittee Meeting, Kennedy Space Center. Chemical Propulsion Information Agency, Columbia, Md. 25 pp.

Lyons, W.A., J.L. Eastman, R.A. Pielke, C.J. Tremback, D.A. Moon, and N.R. Lincoln. 1991b. The mesometeorology of ozone episodes in the lower Lake Michigan air quality region. Paper 91-67, 84th Annual Meeting, Air & Waste Management Association, Vancouver, Canada. 16 pp.

Lyons, W.A., R.L. Walko, M.E. Nicholls, R.A. Pielke, W.R. Cotton, and C.S. Keen. 1992. Observational and numerical modeling investigations of Florida thunderstorms generated by multi-scale surface thermal forcing. Pp. 85-90 in Fifth Conference on Mesoscale Processes, Atlanta. American Meteorological Society, Boston.

Macklin, S.A., G.M. Lackmann, and J. Gray. 1988. Offshore-directed winds in the vicinity of Prince William Sound, Alaska. Mon. Weather Rev. 116:1289-1301.

Macklin, S.A., N.A. Bond, and J.P. Walker. 1990. Structure of a low-level jet over lower Cook Inlet, Alaska. Mon. Weather Rev. 118:2568-2578.

Manins, P.C., and B.L. Sawford. 1979. A model of katabatic winds. J. Atmos. Sci. 36:619-630.

Mass, C.F. 1989. Origin of Catalina eddy. Mon. Weather Rev. 117:2406-2436.

Mass, C.F., and M.D. Albright. 1987. Coastal southerlies and alongshore surges of the west coast of North America: Evidence of mesoscale topographically trapped response to synoptic forcing. Mon. Weather Rev. 115:1707-1738.

Mass, C.F., and G.K. Ferber. 1990. Surface pressure perturbations produced by an isolated mesoscale topographic barrier. Part 1: General characteristics and dynamics. Mon. Weather Rev. 118:2579-2596.

Mass, C.F., M.D. Albright, and D.J. Brees. 1986. The onshore surge of marine air into the Pacific Northwest: A coastal region of complex terrain. Mon. Weather Rev. 114:2602-2627.

McCaul, Jr., E.W. 1991. Buoyancy and shear characteristics of hurricane-tornado environments. Mon. Weather Rev. 119:1954-1978.

McMillen, R.T. 1988. An eddy correlation technique with extended applicability to non-simple terrain. Bound. Layer Meteorol. 43:231-245.

Mizzi, A.P., and R.A. Pielke. 1984. A numerical study of the mesoscale atmospheric circulation observed during a coastal upwelling event on 23 August 1972. Part I: Sensitivity studies. Mon. Weather Rev. 112:76-90.

Moeng, C.H. 1984a. A large-eddy simulation model for the study of planetary boundary-layer turbulence. J. Atmos. Sci. 41:2052-2062.

Moeng, C.H. 1984b. Statistics of conservative scalars in the convective boundary layer. J. Atmos. Sci. 41:3162-3169.

Moeng, C.H., and R. Rotunno. 1990. Vertical-velocity skewness in the buoyancy-driven boundary layer. J. Atmos. Sci. 47:1149-1162.

Moeng, C.H., and J.C. Wyngaard. 1989. Evaluation of turbulent transport and dissipation closures in second-order modeling. J. Atmos. Sci. 46:2311-2330.

Moore, B., and A. Björkström. 1986. Pp. 89-108 in The Changing Carbon Cycle— A Global Analysis. J.R. Trabalka and D.E. Reichle (eds.). Springer-Verlag, New York.

Mulhearn, P.J. 1981. On the formation of a stably stratified internal boundary layer by advection of warm air over a cooler sea. Bound. Layer Meteorol. 21:247-254.

Nagata, M., and Y. Ogura. 1991. A modeling case study of interaction between heavy precipitation and a low-level jet over Japan in the Baiu season. Mon. Weather Rev. 119:1309-1336.

Nappo, C.J., and K.S. Rao. 1987. A model study of pure katabatic flows. Tellus 39A:61-71.

Nelson, C.S. 1977. Wind Stress and Wind Stress Curl over the California Current. NOAA Tech. Rept. NMFS SSRF-714. 87 pp.

Nicholls, S. 1984. The dynamics of stratocumulus: Aircraft observations and comparisons with a mixed layer model. Q. J. R. Meteorol. Soc. 110:783-820.

Nickerson, E.C., and M.A. Dias. 1981. On the existence of atmospheric vortices downwind of Hawaii during the HAMEC Project. J. Appl. Meteorol. 20:868-873.

Nielsen, J.W. 1989. The formation of New England coastal fronts. Mon. Weather Rev. 117:1380-1401.

Nieuwstadt, F.T.M. 1984. The turbulent structure of the stable, nocturnal boundary layer. J. Atmos. Sci. 41:2202-2216.

Nieuwstadt, F.T.M., and R.A. Brost. 1986. The decay of convection turbulence. J. Atmos. Sci. 43:532-546.

Novlan, D.J., and W.M. Gray. 1974. Hurricane-spawned tornadoes. Mon. Weather Rev. 102:476-488.

Ohara, T., I. Uno, and S. Wakamatsu. 1989. Observed structure of a land breeze head in the Tokyo metropolitan area. J. Appl. Meteorol. 28:693-704.

Ohata, T., K. Kobayashi, N. Ishikawa, and S. Kawaguchi. 1985. Structure of the katabatic winds at Mizuho Station, East Antarctica. J. Geophys. Res. 90:10651-10658.

Økland, H. 1990. The dynamics of coastal troughs and coastal fronts. Tellus 42A:444-462.

Olivier, L.D., J.M. Intrieri, and R.M. Banta. 1991. Doppler lidar observations of a land/sea breeze transition on a day with offshore flow. Pp. 138-142 in Proceedings, Fifth Conference on the Meteorology and Oceanography of the Coastal Zone, May 6-9, Miami, Fla. American Meteorological Society, Boston.

Overland, J.E. 1984. Scale analysis of marine winds in straits and along mountainous coasts. Mon. Weather Rev. 112:2530-2534.

Overland, J.E., and P.S. Guest. 1991. The Arctic snow and air temperature budget over sea ice during winter. J. Geophys. Res. 96:4651-4662.

Overland, J.E., and B.A. Walter. 1981. Gap winds in the Strait of Juan de Fuca. Mon. Weather Rev. 109:2221-2233.

Overland, J.E., R.M. Reynolds, and C.H. Pease. 1983. A model of the atmospheric boundary layer over the marginal ice zone. J. Geophys. Res. 88:2836-2840.

Parish, T.R. 1982. Barrier winds along the Sierra Nevada Mountains. J. Appl. Meteorol. 21:925-930.

Parish, T.R. 1984. A numerical study of strong katabatic winds over Antarctica. Mon. Weather Rev. 112:545-554.

Parish, T.R., and D.H. Bromwich. 1989. Instrumented aircraft observations of the katabatic wind regime near Terra Nova Bay. Mon. Weather Rev. 117:1570-1585.

Parish, T.R., and K.T. Waight. 1987. The forcing of Antarctic katabatic winds. Mon. Weather Rev. 115:2214-2226.

Passarelli, Jr., R.E., and R.R. Braham, Jr. 1981. The role of the winter land breeze in the formation of Great Lakes snow storms. Bull. Am. Meteorol. Soc. 62:482-491.

Pielke, R.A. 1984. Mesoscale Meteorological Modeling. Academic Press, New York. 612 pp.

Pielke, R.A. 1989. The status of subregional and mesoscale models for air quality and meteorology: Volume 2. Mesoscale meteorological models in the United States. Environment, Health, and Safety Report 009 5-90 4, December. Pacific Gas and Electric Company, San Ramon, Calif.

Pielke, R.A. 1990. The Hurricane. Routledge Press, London, England. 229 pp.

Pielke, R.A., G. Kallos, and M. Segal. 1989. Horizontal resolution needs for adequate lower tropospheric profiling involved with thermally forced atmospheric systems. J. Atmos. Oceanic Tech. 6:741-758.

Pielke, R.A., G. Dalu, J.S. Snook, T.J. Lee, and T.G.F. Kittel. 1991. Nonlinear influence of mesoscale land use on weather and climate. J. Climate 4:1053-1069.

Pierrehumbert, R.T., and B. Wyman. 1985. Upstream effects of mesoscale mountains. J. Atmos. Sci. 42:977-1003.

Priestly, C.H.B., and R.J. Taylor. 1972. On the assessment of surface heat flux and evaporation using large-scale parameters. Mon. Weather Rev. 100:81-102.

Queney, P. 1948. The problem of airflow over mountains. A summary of theoretical studies. Bull. Am. Meteorol. Soc. 29:16-26.

Raman, S. (ed.). 1982. Proceedings of the Workshop on Coastal Atmospheric Transport Processes. Brookhaven National Laboratory, BNL 51666, UC-11. U.S. Department of Energy, Washington, D.C. 43 pp.

Ramanathan, V., R.D. Cess, E.F. Harrison, P. Minnis, B.R. Barkstrom, E. Ahmad, and D. Hartman. 1989. Cloud-radiative forcing and climate: Results from the Earth Radiation Budget Experiment. Science 243:57-63.

Randall, D.A., J.A. Coakley, C.W. Fairall, R.A. Kropfli, and D.H. Lenschow. 1984. Outlook for research on subtropical marine stratiform clouds. Bull. Am. Meteorol. Soc. 65:1290-1301.

Reason, C.J.C., and M.R. Jury. 1990. On the generation and propagation of the southern African coastal low. Q. J. R. Meteorol. Soc. 116:1133-1151.

Reason, C.J.C., and D.G. Steyn. 1990. Coastally trapped disturbances in the lower atmosphere: Dynamic commonalities and geographic diversity. Progr. in Phys. Geog. 14:178-198.

Reed, R.J. 1980. Destructive winds caused by an orographically induced mesoscale cyclone. Bull. Am. Meteorol. Soc. 61:1346-1355.

Reed, R.J., and C.N. Duncan. 1987. Baroclinic instability as a mechanism for the serial development of polar lows: A case study. Tellus 39:376-384.

Reynolds, M. 1983. Occurrence and structure of mesoscale fronts and cyclones near Icy Bay, Alaska. Mon. Weather Rev. 111:1938-1948.

Reynolds, R.M. 1984. On the local meteorology at the marginal ice zone of the Bering Sea. J. Geophys. Res. 89:6515-6524.

Riordan, A.J. 1990. Examination of the mesoscale features of the GALE coastal front of 24-25 January 1986. Mon. Weather Rev. 118:258-282.

Roeloffzen, J.C., W.D. Van Den Berg, and J. Oerlemans. 1986. Frictional convergence at coastlines. Tellus 38A:397-411.

Rogers, D.P., and D. Koracin. 1992. Radiative transfer and turbulence in the cloud-topped marine atmospheric boundary layer. J. Atmos. Sci. (accepted for publication in June).

Rogers, D.P., and L.M. Olsen. 1990. The diurnal variability of marine stratocumulus clouds. Preprints of the 1990 Conference on Cloud Physics, July 23-27, San Francisco, Calif. American Meteorological Society, Boston.

Rotunno, R. 1983. On the linear theory of land and sea breeze. J. Atmos. Sci. 40:1999-2009.

Samuelsen, R.M. 1992. Supercritical marine layer flow along a smoothly varying coastline. J. Atmos. Sci. (in press).

Sarmiento, J.L., J.R. Toggweiler, and R. Najjer. 1988. Ocean carbon-cycle dynamics and atmospheric P_{CO_2}. In Tracers in the Ocean. Proceedings of a Royal Society Discussion Meeting. H. Charnock, J.E. Lovelock, P.S. Liss, and M. Whitfield (eds.). Royal Society, London.

Schnell, R.C., R.G. Barry, M.W. Miles, E.L. Andreas, L.F. Radke, C.A. Brock, M.P. McCormick, and J.L. Moore. 1989. Lidar detection of leads in Arctic sea ice. Nature 339:530-532.

Schuepp, P.H., M.Y. LeClerc, J.I. MacPherson, and R.L. DesJardins. 1990. Footprint prediction of scalar fluxes from analytical solutions of the diffusion equation. Bound. Layer Meteorol. 50:355-373.

Schuman, U. 1987. Influence of mesoscale orography on idealized cold fronts. J. Atmos. Sci. 44:3423-3441.

Segal, M., R. Avissar, M.C. McCumber, and R.A. Pielke. 1988. Evaluation of vegetation effects on the generation and modification of mesoscale circulations. J. Atmos. Sci. 45:2268-2292.

Sellers, P., Y. Mintz, Y.C. Sud, and A. Dalcher. 1986. A simple biosphere model (SiB) for use within general circulation models. J. Atmos. Sci. 43:505-531.

Sha, W., T. Kawamura, and H. Ueda. 1991. A numerical study on sea/land breezes as a gravity current: Kelvin-Helmholtz billows and inland penetration of the sea-breeze front. J. Atmos. Sci. 48:1649-1665.

Shapiro, M.A., L.S. Fedor, and T. Hampel. 1987. Research aircraft measurements within a polar low over the Norwegian Sea. Tellus 37:272-306.

Shearer, D.L., and R.J. Kaleel. 1982. Critical Review of Studies on Atmospheric Dispersion in Coastal Regions. Nuclear Regulatory Commission, NUREG/CR-2754, Washington, D.C.

Simpson, J.E. 1982. Gravity currents in the laboratory, atmosphere and ocean. Ann. Rev. Fluid Mech. 14:213-234.

Skupniewicz, C.E., J.W. Glendening, and R.F. Kamada. 1991. Boundary layer transition across a stratocumulus cloud edge in a coastal zone. Mon. Weather Rev. 119:2337-2357.

Smith, R.B. 1979. The influence of mountains on the atmosphere. Advances in Geophysics 21:187-230.

Smith, R.B. 1981. An alternative explanation for the destruction of the Hood Canal Bridge. Bull. Am. Meteorol. Soc. 62:1319-1320.

Smith, R.B. 1989. Hydrostatic airflow over mountains. Advances in Geophysics 31:1-41.

Smith, R.K., R.N. Ridley, M.A. Page, J.T. Steiner, and A.P. Sturman. 1991. Southerly changes on the east cost of New Zealand. Mon. Weather Rev. 119:1259-1282.

Smith, S.D. 1988. Coefficients for sea surface wind stress, heat flux, and wind profiles as a function of wind speed and temperature. J. Geophys. Res. 93:467-472.

Smith, S.D., and E.P. Jones. 1985. Evidence for wind-pumping of the air-sea gas exchange based on direct measurements of CO_2 fluxes. J. Geophys. Res. 90:869-875.

Smith, S.D., K.B. Katsaros, W.A. Oost, and P.G. Mestayer. 1990a. Two major experiments in the Humidity Exchange Over the Sea (HEXOS) Program. Bull. Am. Meteorol. Soc. 71:161-172.

Smith, S.D., R.D. Muench, and C.H. Pease. 1990b. Polynyas and leads: An overview of physical processes and environment. J. Geophys. Res. 95:9461-9479.

Smith, S.D., R.J. Anderson, E.P. Jones, R.L. Desjardins, R.M. Moore, O. Herztman, B.D. Johnson. 1991. A new measurement of CO_2 eddy flux in the near shore atmospheric surface layer. J. Geophys. Res. 96:8881-8887.

Smolarkiewicz, P.K., R.M. Rasmussen, and T.L. Clark. 1988. On the dynamics of Hawaiian cloud bands: Island forcing. J. Atmos. Sci. 45:1872-1905.

Smolarkiewicz, P.K., and R. Rotunno. 1989. Low Froude number flow past three-dimensional obstacles. Part I: Baroclinically generated lee vortices. J. Atmos. Sci. 46:1154-1164.

SOS. 1991. Study on Observational Systems: A review of meteorological and oceanographic education in observational techniques and the relationship to national facilities and needs. Bull. Am. Meteorol. Soc. 72:815-826.

Stage, S.A., and J.A. Businger. 1980. A model for entrainment with a cloud-topped marine boundary layer. 1. Model description and application to a cold-air outbreak episode. J. Atmos. Sci. 38:2213-2229.

Stauffer, D.R., N.L. Seaman, and F.S. Binkowski. 1990. Use of four-dimensional data assimilation in a limited-area mesoscale model. Part II: Effects of data assimilation within the planetary boundary layer. Mon. Weather Rev. 119:734-754.

Steyn, D.G., and T.R. Oke. 1982. The depth of the daytime mixed layer at two coastal sites: A model and its validation. Bound. Layer Meteorol. 24:161-180.

Stull, R.B. 1988. An Introduction to Boundary Layer Meteorology. Kluwer, Boston. 666 pp.

Stunder, M., and S. Raman. 1985. A comparative evaluation of the coastal internal boundary layer height equations. Bound. Layer Meteorol. 32:177-204.

Stunder, M., and S. Raman. 1986. A statistical evaluation and comparison of coastal point source dispersion models. Atmos. Environ. 20:301-315.

Tesche, T.W. 1991. Evaluating procedures for using numerical meteorological models as input to photochemical models. Preprint, Seventh Joint Conference on Applications of Air Pollution Meteorology. AMS/AWMA, New Orleans. 4 pp.

Tjernstrom, M. 1990. A numerical study of perturbations in stratiform boundary layer cloud fields on the mesoscale. Pp. 35-38 in Proceedings, Ninth Symposium on Turbulence and Diffusion, April 30-May 3, Roskilde, Denmark. American Meteorological Society, Boston.

Twitchell, P.F., E.A. Rasmusson, and K.L. Davidson. 1989. Polar and Arctic Lows. A. Deepak, Hampton, Va.

Uccellini, L.W., R.A. Petersen, K.F. Brill, P.J. Kocin, and J.J. Tuccillo. 1987. Synergistic interactions between an upper-level jet streak and diabatic processes that influence the development of a low-level jet and a secondary coastal cyclone. Mon. Weather Rev. 115:2227-2261.

Uliasz, M., and R.A. Pielke. 1990. Application of the receptor oriented approach in mesoscale dispersion modeling. Proceedings, Eighteenth NATO/CCMS International Technical Meeting on Air Pollution Modeling and Its Applications, Vancouver, Canada.

Ulrickson, B.L., and C. Mass. 1990a. Numerical investigation of mesoscale circulations in the Los Angeles basin: A verification study. Mon. Weather Rev. 118:2340-2356.

Ulrickson, B.L., and C. Mass. 1990b. Numerical investigation of mesoscale circulations and pollution transport in the Los Angeles basin. Mon. Weather Rev. 118:2357-2401.

Uno, I., H. Ueda, and S. Wakamatsu. 1989. Numerical modeling of the nocturnal urban boundary layer. Bound. Layer Meteorol. 28:693-704.

Venkatram, A. 1986. An examination of methods to estimate the height of the coastal internal boundary layer. Bound. Layer Meteorol. 36:149-156.

Wadhams, P. 1986. The seasonal sea ice zone. In Geophysics of Sea Ice. N. Untersteiner (ed.). Plenum, New York.

Wai, N.M.-K., and S.A. Stage. 1990. Influence of mesoscale sea surface temperature on the break-up of boundary-layer clouds. Pp. 39-42 in Proceedings, Ninth Symposium on Turbulence and Diffusion, April 30-May 3, Roskilde, Denmark. American Meteorological Society, Boston.

Wald, L., and D. Georgopolous. 1984. Atmospheric lee waves in the Aegean Sea and their possible influence on the sea surface. Bound. Layer Meteorol. 28:309-315.

Walko, R.L., W.R. Cotton, and R.A. Pielke. 1992. Large-eddy simulations of the effects of hilly terrain on the convective boundary layer. Bound. Layer Meteorol. 58:133-150.

Walter, B.A., and J.E. Overland. 1982. Response of stratified flow in the lee of the Olympic Mountains. Mon. Weather Rev. 110:1458-1473.

Weil, J.C. 1990. A diagnosis of the asymmetry in top-down and bottom-up diffusion using a Lagrangian model. J. Atmos. Sci. 47:501-515.

Wesley, M.L., D.R. Cook, R.L. Hart, and R.M. Williams. 1982. Air-sea exchange of CO_2 and evidence for enhanced upward fluxes. J. Geophys. Res. 87:8827-8832.

Willis, G.E., and J.W. Deardorff. 1974. A laboratory model of the unstable planetary boundary layer. J. Atmos. Sci. 31:1297-1307.

Winant, C.D., R.C. Beardsley, and R.E. Davis. 1988. Moored wind, temperature and current observations made during Coastal Ocean Dynamics Experiments 1 and 2 over the northern California shelf and upper slope. J. Geophys Res. 92:1569-1604.

Winant, C.D., C.E. Dorman, C.A. Friehe, and R.C. Beardsley. 1988. The marine layer off northern California: An example of supercritical channel flow. J. Atmos. Sci. 45:3588-3605.

Wyngaard, J.C. 1973. On surface-layer turbulence. Pp. 101-149 in Workshop on Micrometeorology. D.A. Haugen (ed.). American Meteorological Society, Boston.

Wyngaard, J.C. 1988. Structure of the PBL. In Lectures on Air Pollution Modeling. A. Venkatram and J. Wyngaard (eds.). American Meteorological Society, Boston.

Wyngaard, J.C., and R.A. Brost. 1984. Top-down and bottom-up diffusion of a scalar in the convective boundary layer. J. Atmos. Sci. 41:102-112.

Wyngaard, J.C., and M.A. LeMone. 1980. Behavior of the refractive index structure parameter in the entraining convective boundary layer. J. Atmos. Sci. 35:1573-1585.

Xu, Q. 1990. A theoretical study of cold air damming. J. Atmos. Sci. 47:2969-2985.

Zanetti, P. 1990. Air Pollution Modeling: Theories, Computational Methods and Available Software. Computational Mechanics Publishers, Boston. 444 pp.

Zeman, O., and J.L. Lumley. 1979. Buoyancy effects in entraining turbulent boundary layers: A second-order closure study. Pp. 295-306 in Turbulent Shear Flows. F. Durst, B.E. Launder, F.W. Schmidt, and J.H. Whitelaw (eds.). Springer-Verlag, Berlin.

Zemba, J., and C.A. Friehe. 1987. The marine atmospheric boundary layer jet in the coastal ocean dynamics experiment. J. Geophys. Res. 92:1489-1496.